The Endogenous Energy-Saving Technological Change in China's Industrial Sector

Xubo He

The Endogenous Energy-Saving Technological Change in China's Industrial Sector

SSAP

Springer

Xubo He
School of Economics
Yunnan University
Kunming, Yunnan, China

Translated by
Jianwei Xiong
Beijing, China

ISBN 978-981-19-7484-7 ISBN 978-981-19-7485-4 (eBook)
https://doi.org/10.1007/978-981-19-7485-4

Jointly published with Social Sciences Academic Press
The print edition is not for sale in China (Mainland). Customers from China (Mainland) please order the print book from: Social Sciences Academic Press.

This Springer imprint is published by the registered company Springer Nature Singapore Pte Ltd.
The registered company address is: 152 Beach Road, #21-01/04 Gateway East, Singapore 189721, Singapore

Preface

Decarbonization of the Chinese economy largely depends on the degree of decarbonization of industry. Since improving energy efficiency and increasing energy R&D investment are arguably the chief drivers of the sustainable growth of the Chinese industry, this book attempts to study energy efficiency and energy R&D investment under one normalized economic model. By distinguishing between energy R&D investment and non-energy R&D investment and drawing on the paradigm of neoclassical economics, I start by clarifying the basic concept of energy-saving technological change (hereinafter referred to as "ESTC") and the related endogenous mechanism, proceed to integrate under the existing endogenous growth model two different mechanisms whereby R&D heterogeneity affects energy efficiency, namely the effect of factor substitution and the effect of increased effective energy input, and finally build a framework of analysis and a mathematical model of R&D heterogeneity-based endogenous ESTC in industry. I analyze the model with the tools of competitive equilibrium, centrally planned equilibrium, and numerical simulation and empirically test the major assumptions of the model with industrial statistics and econometric analytical tools, namely the mixed data regression model, the panel data regression model, and the dynamic panel data regression model.

This book focuses on how R&D heterogeneity influences endogenous ESTC in China's industrial sectors and endeavors to answer the following questions: How does R&D heterogeneity bring about endogenous ESTC? How are the rates of ESTC in the industrial sectors estimated? What is the impact of R&D heterogeneity on the rates of ESTC in the industrial sectors? This book sets itself apart from other existing research by the following work and potentially innovative points.

Firstly, it has constructed an analytical framework for R&D heterogeneity-based endogenous ESTC in industry. Inspired by the paradigm of neoclassical economics, I introduce R&D heterogeneity into the theoretical framework for endogenous growth by distinguishing between energy R&D investment and non-energy R&D investment and differentiating the two mechanisms whereby R&D heterogeneity affects energy efficiency, namely the effect of factor substitution and the effect of increased effective energy input. I integrate these two mechanisms into one framework for discussion, resulting in an analytical framework for R&D heterogeneity-based endogenous ESTC in industry.

Secondly, it has built a mathematical model of R&D heterogeneity-based endogenous ESTC in industry. With the tools of competitive equilibrium, centrally planned equilibrium, and numerical simulation, I analyze the mechanism of R&D heterogeneity-based endogenous ESTC when the elasticity of factor substitution is equal to 1 and when it is not, respectively. I find that (1) both energy R&D investment and non-energy R&D investment boost long-term growth in a competitive equilibrium, though only energy R&D investment favors long-term growth in a planned equilibrium; (2) a larger substitution elasticity leads to a greater rate of ESTC; (3) with energy share constant, improvement of capital quality will reduce (raise) the rate of ESTC if the substitution elasticity is greater than (less than) 1; (4) with capital quality constant, a larger energy share will lower the rate of ESTC; and (5) when the sum of substitution elasticity and the rate of capital improvement is less than (greater than) 1, a larger energy share will increase (decrease) the rate of ESTC.

Thirdly, I have estimated the rates of ESTC in China's 37 industrial sectors from 1980 to 2010. By choosing the best technique for estimating multi-factor CES production functions through comparison, and compiling and applying the quantity- and price-based input-output databases on the 37 industrial sectors from 1980 to 2010, I for the first time estimate the rates of factor-augmenting technological change and elasticities of factor substitution in those sectors under a three-factor CES production function and creatively calculate their rates of ESTC. I find that (1) the (KE) L functional form may fit well with the actual conditions of China's industrial sectors, where the capital-energy substitution elasticity is estimated to be between 0.023 and 1.028; the composite-labor substitution elasticity is estimated to be between 0.122 and 1.088; and the energy-, capital-, and labor-augmenting technological change rates are estimated to be between 0.008 and 0.431, between −0.015 and −0.234, and between 0.023 and 0.14, respectively; and (2) ESTC is requisite for reducing energy intensities in China's industrial sectors. The rates of ESTC in high energy-consuming sectors and energy sectors are higher than in low energy-consuming sectors and non-energy sectors, respectively, exhibiting positive U-shaped curves from 1980 to 2010.

Fourthly, I have empirically tested the relationship between R&D heterogeneity in China's industrial sectors and the rates of ESTC. In this book, I measure R&D heterogeneity on the basis of the R&D intensity, R&D scale, and characteristics of China's industrial sectors and empirically test the relationship between R&D heterogeneity and the rates of ESTC with the current mainstream econometric analytical tools, namely the mixed data regression model, the panel data regression model, and the dynamic panel data regression model, using statistics on the industrial sectors from 2000 to 2010. I find that R&D heterogeneity in China's industrial sectors can significantly improve ESTC. Specifically, (1) R&D intensity and R&D scale have greater improving effects on ESTC in high energy-consuming sectors and energy sectors than in low energy-consuming sectors and non-energy sectors; (2) while R&D intensity has comparable improving effects on ESTC in high energy-consuming sectors and energy sectors, R&D scale has greater improving effects on ESTC in high energy-consuming sectors than in energy sectors.

Kunming, China

Xubo He

Contents

Chapter 1
Introduction

1.1 Research Background and the Question Raised

1.1.1 Research Background

Energy is the foundation and guarantee of economic growth and social development. Since the introduction of the reform and opening up policy in the late 1970s, China has experienced rapid economic growth, which has lifted large numbers of people out of poverty, but meanwhile has produced tremendous pressure on the energy market and the natural environment. According to statistics provided by British Petroleum (BP), China accounted for 20.3% of the global primary energy consumption in 2010, including 10.6% of oil, 3.4% of natural gas, and 48.2% of coal. In the past 30 years, Chinese industry contributed only 40.1% of GDP, but consumed 67.9% of energy and emitted 83.1% of carbon dioxide nationwide (Shiyi, 2009). With urbanization and industrialization accelerating, China's reliance on energy will only increase in the future. Owing to the scarcity of energy (especially fossil fuels) and the greenhouse effect caused by the combustion of fossil fuels, energy constraints will have increasing impacts on economic growth, the environment and industrial transformation for an extended period, posing the following challenges.

(1) Energy Security Concerns

Following the inauguration of the reform and opening up policy, China's total energy production increased from 627.7 million tonnes of coal equivalent (TCE) in 1979 to 2,969.16 million TCEs in 2010, while its total energy consumption jumped from 571.44 million TCEs to 3,249.32 million TCEs over the same period. According to the *China Statistical Yearbook*, a gap between energy supply and demand emerged in 1992, which was about 20 million TCEs in that year and increased year by year thereafter. The demand–supply gap not only picked up with years in oil and coal, but also appeared in 2007 in natural gas, which had a small proportion in the energy consumption structure. Table 1.1 shows the supply–demand gaps in China's total energy and other major fossil fuels in every five years from 1990 to 2010. As shown

X. He, *The Endogenous Energy-Saving Technological Change in China's Industrial Sector*, https://doi.org/10.1007/978-981-19-7485-4_1

Table 1.1 Energy supply–demand gaps in China from 1990 to 2010

Year	Total gap	Gap in coal	Gap in oil	Gap in natural gas
1990	5219	1898.438	3360.482	5.677
1995	−2142	−694.694	−1536.16	90.478
2000	−10,483	−1852.32	−9079.63	444.614
2005	−19,778	700.068	−20,781.1	350.648
2010	−28,023	6479.136	−32,640.6	−1826.84

Source China Statistical Yearbook 2012
Note The figures are obtained by deducting consumption from the production of each energy source in 10,000 tonnes of coal equivalent

in the table, while the supply–demand gap in oil is biggest at present, those in raw coal and natural gas are likely to widen with China's dependence on the availability of energy rising for development. In 2010, the gap between supply and demand in natural gas stood at 18.2684 million TCEs. It is most likely, therefore, that China's future energy security hinges not only on the stable supply of oil, but also on all other fossil fuels. In addition to energy security concerns, there may arise a situation where China, in order to avoid the risks of geopolitical changes and price fluctuations in the international oil and gas markets, has to live with a coal-based single energy source consumption structure (Yuyan & Qingyou, 2007).

(2) Low Energy Efficiency

While China's energy supply depends heavily on imports, the efficiency in energy production and consumption is very low. In the existing literature, energy consumption per unit of GDP, i.e., energy intensity, is generally used as a measure of energy efficiency (Chu & Manhong, 2007a, b, 2009). Table 1.2 shows energy intensity in China and other major regions in the world in 2000 and 2010. As shown in the table, at the 2005 exchange rate, China's energy intensity was 0.79 tonne of oil equivalent (TOE) per thousand US dollars of output in 2000, compared with the global average of about 0.24 TOE, and a figure as low as 0.16 TOE in OECD countries, which means that China's energy efficiency was only 1/5 of the OECD average. Under the current rate method, China's energy intensity was higher than that in other parts of the world in 2000, but was only 0.24 TOE calculated by the PPP method, slightly higher than the average levels of OECD countries and Asia, and equal to the world average. In 2010, China's energy intensity was 0.64 TOE by the current rate method, about 1/5 lower than that in 2000, but much higher than the average of 0.15 TOE in OECD countries. In 2010, China's energy intensity was 0.27 TOE under the PPP method, a great improvement on that under the current rate method. Nevertheless, it increased over 2000. In the same duration, energy intensity in regions other than China and non-OECD Europe declined in varied degrees. By the PPP method, energy intensity in China was slightly lower than in non-OECD Europe in 2010, but was higher than the world average as well as the OECD average.

Table 1.2 Energy intensity in China and other major regions in the world

	2000		2010	
	By current rate method	By PPP method	By current rate method	By PPP method
World	0.24	0.24	0.25	0.19
OECD	0.16	0.22	0.15	0.15
Middle East	0.54	0.4	0.51	0.26
Non-OECD Europe	0.59	0.28	0.74	0.32
China	0.79	0.24	0.64	0.27
Asia	0.54	0.23	0.49	0.18
Latin America	0.24	0.18	0.27	0.14
Africa	0.72	0.33	0.56	0.25

Sources IEA: Key World Energy Statistics 2002 & 2012; BEA: Current Dollar and Real GDP
Note Energy intensity under the current rate method is calculated using the 2005 US dollar exchange rate. All numbers are presented in TOE per thousand US dollars of output

(3) Huge Pressure to Address Climate Change

The coal-based energy consumption structure (Fig. 1.1) and low energy efficiency have combined to result in tremendous carbon emissions in China (Fig. 1.2). As carbon emissions are difficult to measure directly, international and domestic scholars tend to use the accounting method developed by the Intergovernmental Panel on Climate Change (IPCC) for the United Nations Framework Convention on Climate Change (UNFCCC) in 2006 to estimate carbon emissions. Estimations vary due to different underlying data sources. BP statistics show that China had become the world's largest carbon emitter up to 2006, while research of the International Energy Agency (IEA) ranks China as the number one carbon emitter in the world since 2009. International Energy Agency. World Energy Outlook 2010[M]. Paris: IEA, 2010. The National Energy Administration of China, however, contends that both BP and IEA overestimate China's carbon emissions. While different levels of emissions correspond to different responsibilities and political interests for China, it is indisputable that, according to statistics published by the *China Statistical Yearbook*s, China's coal consumption has increased rapidly since 2002. It is precisely because of coal consumption greatly outpacing oil and gas consumption that China's carbon emissions have soared rapidly in the recent decade.[1] China was not committed to reduce emissions under the Kyoto Protocol; but, in order to cultivate the image of a "responsible major country", the Chinese government has taken actions to save energy and

[1] Among the carbon dioxide emission factors recommended by IPCC, EIA, the Chinese Academy of Engineering and the Asian Development Bank, all of the coal emission factors are bigger than the oil and gas emission factors. See http://xmecc.xmsme.gov.cn/2010-11/20101116111816.htm.

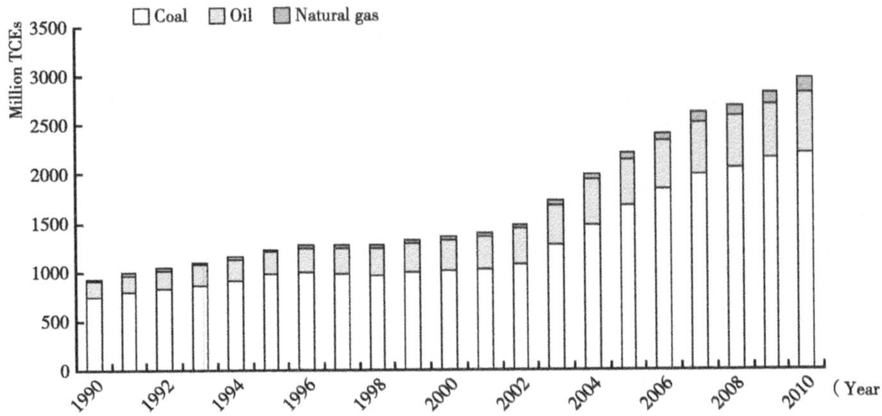

Fig. 1.1 Fossil fuels consumption structure of China 1990–2010. *Source* China Statistical Yearbook 2011

control carbon emissions. In 1997, the Standing Committee of the National People's Congress enacted the Energy Conservation Law, with energy conservation becoming a basic national policy. The 11th Five-Year (2006–2010) Plan promulgated in 2006 made it a development goal for the first time to reduce energy consumption per unit of GDP. In 2009, the Chinese government officially announced its goal of controlling GHG emissions at the Copenhagen Global Climate Conference—a 40–45% reduction in carbon emissions per unit of GDP in 2020 from the 2005 levels. In 2011, the State Council set the target in its Work Plan for Controlling GHG Emissions during the 12th Five-Year Plan Period (2011–2015): by 2015, emissions would be 17% lower than in 2010. With increasing pressure on China to cope with climate change, it is of great practical significance to study how to reduce carbon emissions and honor international commitments without compromising long-term economic growth.

(4) Acute Energy and Emissions Reduction Challenges Facing Industry

Relative to the Chinese economy as a whole, the industrial sectors are more reliant on the supply of energy. Figure 1.3 shows the proportions of China's industrial output and industrial energy consumption in China's GDP and total energy consumption respectively. As shown in the figure, industry accounted for 65% of the country's energy consumption and contributed 37% of GDP in 1990. The industrial growth characterized by high energy consumption moderated around 2000. In 2001, the industrial sectors produced 50% of GDP and consumed 61% of the country's delivered energy, a significant improvement in energy efficiency over 1990. After 2005, however, energy efficiency in industry began to deteriorate. In 2010, the industrial sectors used 65% of energy but produced only 50% of GDP. Figure 1.4 compares the energy intensity of the industrial sectors and of the whole country from 1990 to 2010, measured by TCE per RMB 100 of output in 1990. As shown in Fig. 1.4, the energy intensity of Chinese industry decreased from 9.37 TCEs in 1990 to 3.15 TCEs in

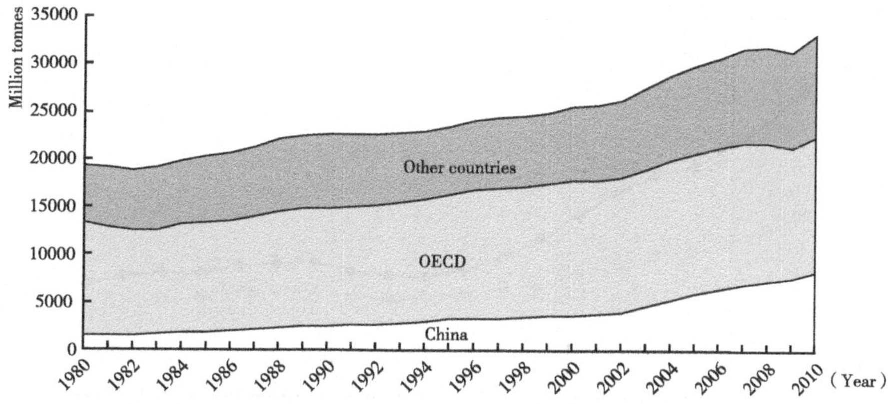

Fig. 1.2 1980–2010 global carbon dioxide emissions, *Source* BP (2011)

2010, while the energy intensity of the whole country fell from 5.32 TCEs in 1990 to 2.39 TCEs in 2010. Although the industrial sectors' energy intensity declined steeply from 1990 to 2010, it was still higher than the national average.

The energy intensity of Chinese industry dropped significantly from 1980 to 2010. How was it distributed across the industrial sectors? How did that distribution evolve from 1980 to 2010? Fig. 1.5 shows the distribution and evolution of energy intensity in the 2-digit Standard Industrial Classification (SIC) sectors in every ten years. As indicated in the figure, industrial energy intensity decreased on average from 1980 to 2010 by more than 50%, from over 10 million TCEs per RMB 100 million of value added in 1990 (Fig. 1.5a) to less than 5 million TCEs per RMB 100 million of value

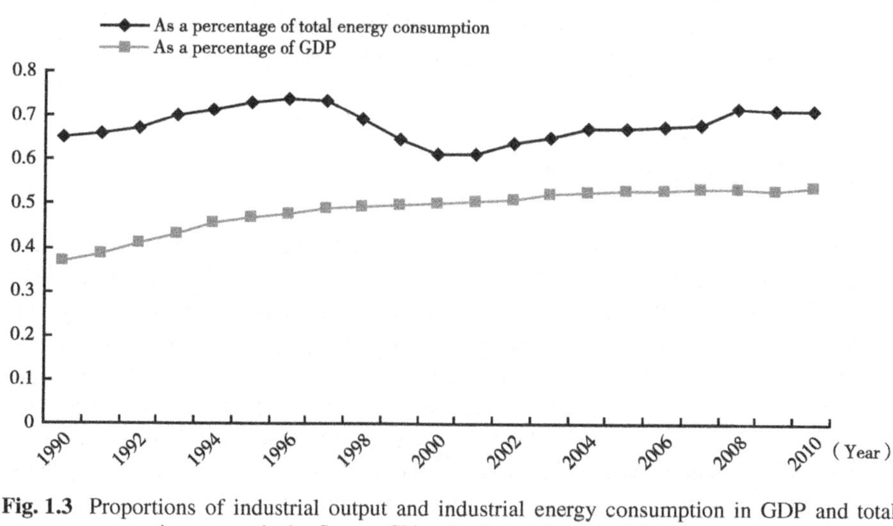

Fig. 1.3 Proportions of industrial output and industrial energy consumption in GDP and total energy consumption respectively. *Source* China Statistical Yearbook 2012. *Note* The decimals in the figure may translate into percentages

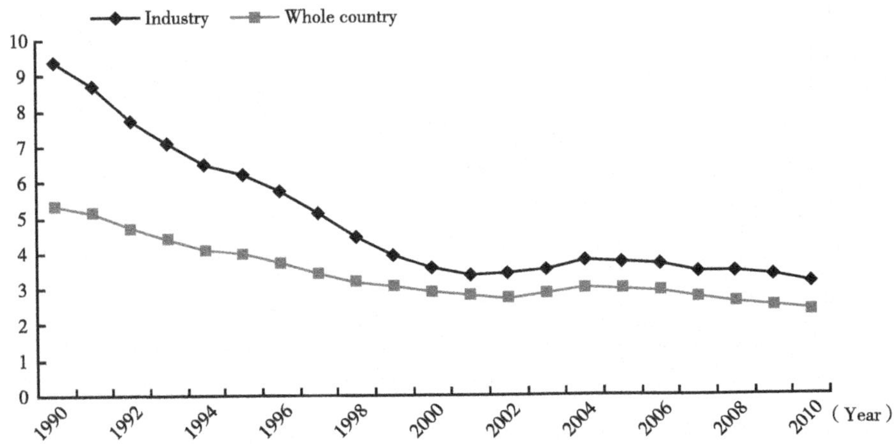

Fig. 1.4 Energy intensity of industry and of the whole country. *Source* China Statistical Yearbook 2012. *Note* The unit is TCE per RMB 100 of output in 1990

added in 1990 (Fig. 1.5d). In addition, in terms of energy intensity distribution across the sectors, kernel density gradually evolved from a double-peak distribution in 1980 to a single-peak distribution in 2000, and shifted to the left during the period, meaning that energy intensity in most sectors declined significantly. Although a double-peak distribution of energy intensity appeared again in 2010, the density gap between the two peaks was much larger than in 1980, and the peak value and average value in 2010 moved to the left more than in 2000. Finally, despite large energy intensity decreases in most sectors, energy intensity spikes were seen in some sectors, exhibiting a "long tail" distribution as shown in Fig. 1.5c, d. It can be inferred from Fig. 1.5 that improving energy intensity in those sectors on the "long tail" should be the key to saving energy and reducing emissions in China's industrial sectors.

1.1.2 The Question Raised

It follows from the above analysis that subject to the constraints of energy security and climate change, China's sustainable growth is predicated on and guaranteed by the low-carbon transition of the industrial sectors, which in turn depends on reducing the intensity of fossil fuels used currently in production processes, especially in those sectors on the "long tail", to enable more efficient utilization of energy. To attain this goal, the existing literature suggests three approaches (Popp et al., 2009; Zhou and Tamas, 2010; Liu et al., 2011): (1) improving the efficiency in exploiting existing energy sources (especially fossil fuels); (2) increasing the share of clean energy in the structure of industrial energy consumption; and (3) increasing energy- and climate-related R&D investment in industry. As carbon emissions are chiefly caused by the burning of fossil fuels, all of the approaches are effective at reducing energy use

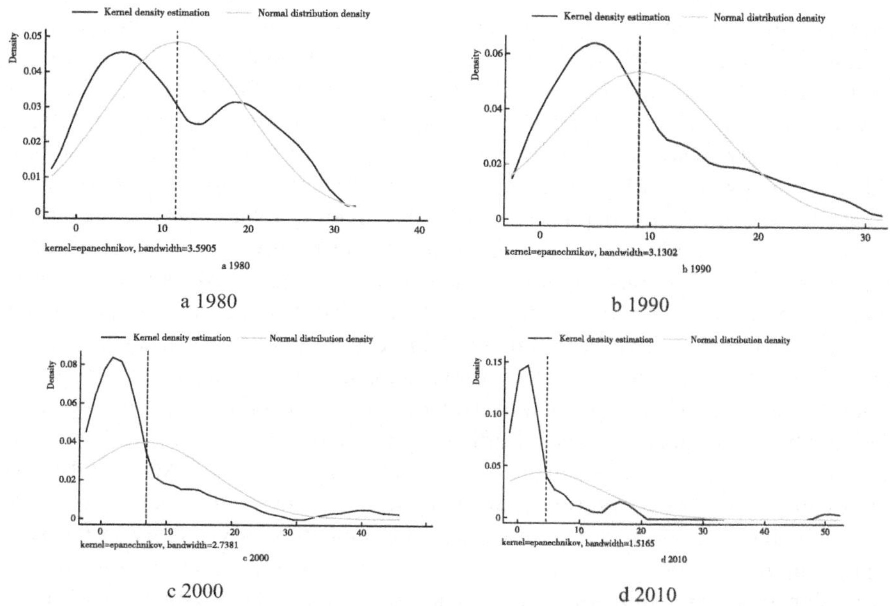

Fig. 1.5 Energy intensity distribution and evolution in China's industrial sectors: 1980–2010. *Source* Raw data on output and total energy consumption by sector come from the *China Statistical Yearbook*. How the data are processed is described in Chap. 5 of this book. *Note* The X-axis in the figure represents the value of energy intensity in million TCEs per RMB 100 million of value added in 1990. The Y-axis shows sectoral densities corresponding to various values of energy intensity. The vertical line denotes the average energy intensity in all the industrial sectors in a particular year concerned

and emissions. However, thanks to the characteristics of clean energy, including low energy density (e.g., biomass), intermittency (e.g., solar and wind power) and high costs,[2] large-scale, stable supplies of clean energy sources are so difficult to come by in the short run that they are hardly competitive in the marketplace and ready for mass adoption (Jingli, 2010; Gowrisankaran et al., 2011; Cruz & Taylor, 2013). Therefore, improving energy efficiency and increasing energy- and climate-related R&D investment (hereinafter called "energy R&D investment") have figured as the principal pathway to a low-carbon future of Chinese industry under the constraints of energy security and climate change.

How is the efficiency in energy utilization measured? The 14th Conference of the Parties to the UNFCCC held in December 2008 for the first time drew the attention of all Parties to the measurement of energy efficiency. At present, in most of the

[2] Take as an example the fairly mature technology of generating electricity with renewable energy sources in China. At present, the price is RMB 0.8–1.0 per kWh for wind power, RMB 0.8–1.5 per kWh for photovoltaic power, and RMB 2.0 per kWh for solar thermal power generation, 2–5 times higher than the RMB 0.23–0.47 per kWh for coal-fired generation. For information about the reality of generating electricity with renewables and the pricing policy in China, please read Jingli (2010).

relevant literature published at home and abroad, energy efficiency is measured by indicators such as energy intensity (the ratio of energy supply to output), energy productivity (the reciprocal of energy intensity), and total-factor energy efficiency (less energy is used to produce a given quantity of output),[3] but none of them is capable of fully measuring the efficiency in energy utilization (Chu & Manhong, 2007a, b; Qingwen, 2012). Taking energy intensity as an example, energy intensity at the national level is generally calculated by dividing total energy consumption by GDP. By this method, any additional investment in physical or human capital, even with total energy consumption unchanged, will increase GDP, or will entail less energy to produce the same amount of GDP (provided that one factor of production can be substituted for another), thus reducing energy intensity. That is to say, the key to improving energy efficiency in this exercise is to "squeeze out" energy by other non-energy production factors. However, as the exercise does not consider the possibility of improving energy efficiency by increasing effective energy input, energy intensity is not a comprehensive metric of energy efficiency. Also, total-factor energy efficiency has measuring problems. Because total-factor energy efficiency describes the production process of a decision-making entity mainly in terms of input–output changes in engineering science, it may neglect how the quantitative expansion or quality improvement of a non-energy production factor will affect energy efficiency. A full measurement of energy efficiency requires an answer to the following critical question: How should one introduce both the effect of factor substitution and the effect of increased effective energy input into the system of indicators of energy efficiency?

Does energy R&D investment influence energy efficiency? If the answer is yes, then what should we do to include both energy R&D investment and energy efficiency in one economic model for analysis? Since improving energy efficiency and increasing energy R&D investment are the key to the low-carbon transition of China's industrial sectors, is it possible that we discuss both of them in one economic model, and analyze both the effect of factor substitution and the effect of increased effective energy input that impact energy efficiency? As endogenous growth theory has built a full analytical framework for topics ranging from R&D investment to technological change to sustainable growth, how should we construct and analyze a logical chain from energy R&D investment to energy efficiency improvement to sustainable growth based on endogenous growth theory? Scholz and Ziemes (1999), Schou (2000, 2002), Grimaud and Rouge (2003), van Zon and Yetkina (2003), Hassler et al. (2012), and Wen and Zhaoxi (2011, 2012) have created their models based on endogenous ESTC theory to describe and analyze the said logical chain. Whereas Hassler et al. (2012) consider the impact of increased energy input on energy efficiency, other authors focus only on the effect of factor substitution brought about by the quantitative expansion and quality improvement of other non-energy factors (especially the

[3] Most of the domestic literature does not distinguish between energy productivity and energy efficiency. This book adopts the definitions given by Chu and Manhong (2007a, b) and Qingwen (2012) by referring to the reciprocal of energy intensity as energy productivity and to the optimal ability to use energy as energy efficiency. For nuances between the two concepts, read Chu and Manhong (2007a, b), Qingwen (2012) and Chap. 3 of this book.

improvement of capital quality). Although Hassler et al. (2012) consider the possible impact on ESTC of competitive heterogeneous R&D investments—one type of R&D investment improves capital quality while the other favors ESTC, their assumption that capital and energy are fully complementary to each other excludes the effect of factor substitution from consideration. This immensely limits the applicability of their model.

In addition, the existing literature does not touch on the relationship between ESTC and other existing energy efficiency indicators. In fact, compared with energy productivity and energy efficiency, ESTC has two advantages in measuring energy efficiency. Firstly, an optimally-chosen ESTC indicator can measure energy efficiency comprehensively, enriching the existing system of indicators by supplying the defect of the existing energy efficiency indicators that focus either on the effect of factor substitution or on the effect of increased effective energy input. Secondly, starting from the concept of ESTC defined in this book, measuring energy efficiency by ESTC can fully depict the logical chain from energy R&D investment to energy efficiency improvement to sustainable growth. This in turn leads to the following questions: First, how should we measure ESTC? Second, what are the similarities and differences between ESTC and the existing energy efficiency indicators? Third, how is ESTC distributed in China's industrial sectors? Fourth, what is the impact of R&D heterogeneity on ESTC in Chinese industry?

To answer those questions, I first make the existing ESTC framework my theoretical cornerstone, clarify the basic concept and implications of ESTC, and draw on the basic theories and analytical tools of standard neoclassical economics, so as to build an analytical framework for R&D heterogeneity-based endogenous ESTC and a related mathematical model to explain the mechanisms and avenues whereby R&D heterogeneity impacts ESTC. Then, I compare and critique the mainstream methods of estimating constant elasticity of substitution (CES) production functions, and select a suitable one to estimate the rates of ESTC in China's 37 industrial sectors. Finally, I empirically analyze the possible impact of R&D heterogeneity on the rates of ESTC in Chinese industry to verify the core assumptions suggested in this book.

1.2 Research Significance

1.2.1 Theoretical Significance

The theoretical framework built in this book represents a fresh theoretical exploration of the low-carbon transition of China's industrial sectors and sustainable growth. Its theoretical value can be summarized as follows:

(1) It corrects the inadequacy of the existing energy efficiency indicators

Since none of energy intensity, energy productivity, energy efficiency or total-factor energy efficiency is capable of fully measuring how efficiently energy is utilized,

I draw on and expand the concept of ESTC in the existing literature, and further elucidate its definition. By distinguishing between energy R&D investment and non-energy R&D investment, I consider both mechanisms whereby R&D heterogeneity influences energy efficiency. My comparative study of the state-of-the-art research and accomplishments in sustainable growth, climate change and energy economics in China and abroad finds that foreign scholars have had the tradition of differentiating between energy efficiency and ESTC since Hicks (1932). In particular, during the oil crisis in the 1970s and after the conclusion of the Kyoto Protocol in the 1990s, most of the relevant literature discusses the similarities and differences between energy efficiency and ESTC. At present, outside of China, the literature on energy efficiency focuses on microscopic subjects, while the literature on ESTC deals with macroscopic ones (Losche, 2002; Gillingham et al., 2008; Lösche & Schymura, 2013; Zhou et al., 2008). Domestic research on energy efficiency basically compares energy efficiency of different firms, industries and even provinces and analyzes possible causes thereof (Chu & Manhong, 2007a, b), while theoretical and empirical research on ESTC, especially in China's industrial sectors, has remained in limbo.

(2) It puts together a model of R&D heterogeneity-based endogenous ESTC

Most of the existing endogenous ESTC models postulate that energy efficiency can be improved by replacing energy input in the production function with a better non-energy production factor or a greater number of it (especially capital), or through a single mechanism such as increased effective energy input. As both mechanisms are effective at improving energy efficiency as noted above, the existing research cannot, strictly speaking, be called comprehensive endogenous ESTC models. Both energy efficiency and ESTC are abstract ideas that can hardly be proven with existing statistics. Given energy intensity's role in measuring changes in energy efficiency, our priority should be building an endogenous ESTC model with the characteristic that ESTC is at least requisite for improving energy intensity, in other words, ESTC should to some extent explain energy intensity improvements.

(3) It expands the existing endogenous ESTC models

The existing endogenous ESTC models either neglect the possible impact of increased effective energy input on energy efficiency, or fail to consider both the effect of factor substitution and the effect of increased energy input in one framework of analysis, thus significantly limiting their cogency and applicability. This book builds mainly on the work of Aghion and Howitt (1998), Scholz and Ziemes (1999), Schou (2000, 2002), Grimaud and Rouge (2003), Wen and Zhaoxi (2011, 2012), and extends the research boundary of Hassler et al. (2012). The model built in this book loosens the assumption that capital and energy are fully complementary to each other, and analyzes optimal behaviors in competitive equilibrium and centrally planned equilibrium respectively to deduce the best allocation of R&D investment factors.

(4) It provides a brand new perspective for the low-carbon transition of Chinese industry and sustainable growth

The issue of decarbonization of industry and sustainable growth centers on how to resolve energy constraints on economic growth, and how to handle the relationship between the greenhouse effect associated with energy use and sustainable growth. Apart from energy efficiency and energy R&D investment, Chinese and foreign scholars have approached this issue from three other theoretical angles. The first is the environmental Porter Hypothesis (Porter, 1991). Porter believes that high energy consumption and high emissions are signals of some form of economic waste and inefficient use of resources, and curing this inefficiency will give manufacturing sectors the potential for improving environmental quality as well as productivity, a win–win situation where manufacturing sectors can reduce energy use and GHG emissions while sustaining growth. The second is decoupling theory, according to which decoupling refers to the breaking of the link between economic growth and environmental pressures (Jianping et al., 2011). According to OECD (2001), decoupling can be either absolute or relative. The former occurs when environmental pressures are stable or decreasing while the economy is growing. Decoupling is relative when the economy grows with environmental pressures, but faster than the rate of environmental pressures. From the perspective of reducing energy use and emissions, decoupling means that economic growth outpaces carbon emissions or fossil fuel consumption. The third theory is concerned with factor substitution. Assuming that capital and energy are mutually complementary, the theory of factor substitution holds that economic growth and capital expansion and deepening must be accompanied by increased energy consumption (Berndt & Wood, 1975). Chengjun and Ruiming (2008), Xiaoma et al. (2009) and Juan et al. (2010) have contributed typical research in this field. The weakness shared by the three theories described above is that most studies stay at data analysis without delving into the core mechanisms behind the Porter Hypothesis, decoupling or factor substitution theory, let alone distinguishing ways in which energy efficiency is improved or integrating the effect of factor substitution and the effect of increased effective energy input in one framework of analysis.

1.2.2 Practical Significance

According to the *2012 Annual Report on China's Policies and Actions to Address Climate Change* prepared by the Climate Change Department of the National Development and Reform Commission (NDRC), 2011 witnessed nationwide the shutdown of about 8 million kilowatts of small coal-fired generating units and the elimination of outdated production capacities of 31.92 million tonnes of iron, 28.46 million tonnes of steel, 155 million tonnes of cement (clinker and mill), 20.06 million tonnes of coke, 30.41 million weight cases of flat glass, 8.3 million tonnes of paper, 639,000 tonnes of electrolytic aluminum, 425,000 tonnes of copper, 661,000 tonnes of lead, and 48.7 million tonnes of coal. Although shutting down old production capacities in energy-intensity sectors may be more or less effective at reducing carbon emissions, whether it helps improve energy efficiency in China's industrial sectors remains to be seen. As far as the low-carbon transition and sustainable development of Chinese industry is concerned, administrative prescriptions may cure the symptoms of a disease but not the underlying malady. Worse still, implementing energy saving and emissions reduction policies simply through "power cuts" will have adverse effects on economic growth as it may lead to higher prices of raw materials (Boqiang, 2010).

In the wake of the global financial crisis, calls for economic growth began to overwhelm environmental concerns on the ground that implementing measures to reduce energy use and emissions in manufacturing sectors would virtually increase production costs, which would in turn erode market competitiveness. Some Western scholars even claim that it is extravagant to talk about reducing energy use and emissions while the financial crisis is raging (McKinsey, 2010). Research of the Club of Rome suggests that the industrialization path characterized by high levels of investment and output since the Industrial Revolution is unsustainable as it may result in resource shortage and environmental pollution. Does reducing energy use and emissions conflict with economic growth? What should be done to help Chinese industry shift away from the inefficient growth model driven by high energy consumption and emissions to a low-carbon and sustainable one? These questions represent real challenges to China's effort to reduce energy use and emissions and demand prompt and proper solutions, which will help ensure successful implementation of related measures.

1.3 Research Goal and Methodology

1.3.1 Research Goal

The overarching goal of this book is to build a normative and logically consistent framework of analysis to reveal the mechanism of R&D heterogeneity-based endogenous ESTC, and to establish with empirical data the relationship between R&D heterogeneity and the rates of ESTC in China's industrial sectors. It is comprised of four objectives. The first is to build an analytical framework for R&D heterogeneity-based ESTC in the industrial sectors based on the existing research, with aims at improving, expanding and enriching the existing frameworks. The second is to establish a basic and analyzable mathematical model based on the framework to explain how R&D heterogeneity works and flows through. The third objective is to estimate the rates of ESTC in China's industrial sectors with an appropriate measurement technique. The final objective is to empirically test the relationship between R&D heterogeneity and the rates of ESTC in China's industrial sectors.

1.3.2 Methodology

(1) Combination of Empirical Analysis with Normative Analysis

I employ empirical analysis, which consists of equilibrium analysis and numerical simulation to build the ESTC framework, estimate the rates of ESTC in China's industrial sectors, and analyze the impact of R&D heterogeneity on ESTC. In addition, I use an econometric method to empirically test the theoretical logic and basic mathematical empirical conclusions made under the framework. This book ends with policy recommendations on the basis of the findings obtained through normative analysis.

(2) Combination of Theoretical Logic Deduction with Empirical Analysis

Under the theoretical logic deduction method, I use the paradigm of neoclassical economics and the analytical tools of endogenous growth theory to identify the mechanism and ways whereby R&D heterogeneity impacts ESTC. In empirical analysis, I first estimate the rates of ESTC in China's industrial sectors with a three-factor CES production function, and then empirically test the relationship between R&D heterogeneity and the rates of ESTC, i.e., the logically deduced basic propositions, with the modern econometric analytical tools of the mixed regression model, the panel data fixed effect model and the dynamic panel data model.

1.4 Contents and Technical Flow Chart

1.4.1 Contents

This book is made up of seven chapters. This chapter explains the research background, significance, methodology, contents, technical flow chart and potentially innovative points. Chapter 2 contains a review of the relevant theories and literature. Chapters 3 and 4 are devoted to the building of an analytical framework for and a mathematical model of R&D heterogeneity-based endogenous ESTC in the industrial sectors, and a theoretical analysis of the mechanism and avenues through which R&D heterogeneity affects ESTC. Chapter 5 estimates the rates of ESTC in China's industrial sectors. Chapter 6 is an empirical analysis of R&D heterogeneity-based endogenous ESTC in China's industrial sectors. Chapter 7 presents the main findings and outlook. Simply put, this chapter and Chap. 2 introduce the research background; Chaps. 3 and 4 deal with theoretical topics; Chaps. 5 and 6 address empirical issues; and Chap. 7 summarizes the basic findings from the theoretical analysis and empirical research, presenting the central arguments, conclusions and shortcomings of this book.

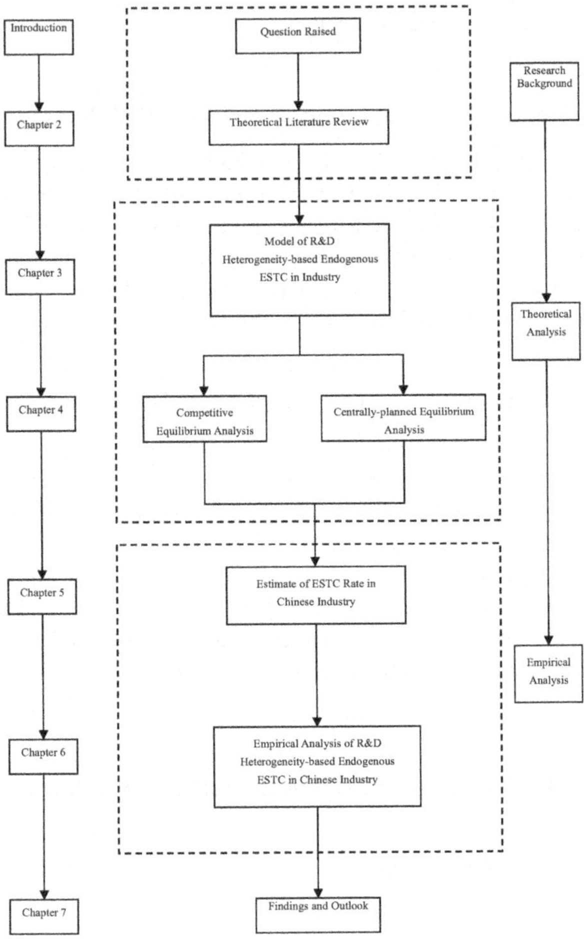

Fig. 1.6 Technical flow chart

1.4.2 Technical Flow Chart

The technical flow chart of this book is shown in Fig. 1.6.

References

Aghion, P., & Howitt, P. (1998). *Endogenous Growth Theory*. MIT Press.

Berndt, E. R., & Wood, D. O. (1975). Technology, price, and the derived demand for energy. *The Review of Economics and Statistics, 57*(3), 259–268.

Boqiang, L. (2010). Power cuts are not a viable pathway to energy conservation. *The Journal of Environmental Economics, 10*, 64.

BP. (2011). *BP Statistical Review of World Energy*. BP.

Chengjun, L., & Ruiming, Z. (2008). A study of energy substitution in the Chinese industrial sector—Based on revision of the ALLEN substitution elasticity model. *The Journal of Quantitative and Technical Economics, 05*, 30–42.

Chu, W., & Manhong, S. (2007a). Energy efficiency and the influencing factors: A DEA-based empirical analysis. *The World of Management, 08*, 66–76.

Chu, W., & Manhong, S. (2007b). Energy efficiency and energy productivity: A comparison of inter-provincial data by DEA method. *Quantitative and Technical Economics, 09*, 110–121.

Chu, W., & Manhong, S. (2009). Developments and trends in energy efficiency research: A summary. *The Journal of Zhejiang University (Humanities and Social Sciences Edition), 03*, 55–63.

Cruz, J. M., & Taylor, M. S. (2013). *A spatial model to energy economics* (NBER Working Paper No. 18908).

Gillingham, K., Newell, R. G., & Pizer, W. A. (2008). Modeling endogenous technological change for climate policy analysis. *Energy Economics, 30*(6), 2734–2753.

Gowrisankaran, G., Reynolds, S. S., & Samano, M. (2011). *Intermittency and the value of renewable energy* (NBER Working Paper No. 17086).

Grimaud, A., & Rouge, L. (2003). Non-renewable resources and growth with vertical innovation optimum, equilibrium and economic policies. *Journal of Environmental Economic and Management, 45*, 433–453.

Hassler, J., Krusell, P., & Olovsson, C. (2012). *Energy-saving technical change* (NBER Working Paper No. 18456).

Hicks, J. R. (1932). *The Theory of Wages*. Macmillan.

International Energy Agency. (2010). *World Energy Outlook 2010*. IEA.

Jianping, Z., Fangfang, T., & Hao, F. (2011). Energy consumption, carbon emissions and industrial growth—An empirical analysis from a decoupling perspective. *Contemporary Economic Science (当代经济科学), 06*, 81–89+125.

Jingli, S. (2010). *China's pricing policy on electricity generated from renewable energy sources [EB/OL]*. National Development and Reform Commission Energy Research Institute. Retrieved from http://www.efchina.org/csepupfiles/report/201112441339800.6199902107537.pdf/4.%20Chinese%20RE%20Pricing%20Policies_SHI%20Jingli_CN.pdf

Juan, G., Chonghui, G., & Ling, Y. (2010). On the energy rebound effect in the Chinese industrial sector. *The Journal of Quantitative and Technical Economics, 11*, 114–126.

Liu, T., Xu, G., Cai, P., et al. (2011). Development forecast of renewable energy power generation in China and its influence on the GHG control strategy of the country. *Renewable Energy, 36*(4), 1284–1292.

Lösche, A., & Schymura, M. (2013). *Modeling technological change in economic models of climate change: A survey* (ZEW Working Paper No. 13-007).

Losche, A. (2002). Technological change in economic models of environmental policy: A survey. *Ecological Economics, 43*(2), 105–126.

McKinsey. (2010). *Connecting climate change and economic recovery [R/OL]*. McKinsey & Company. Retrieved from http://www.mckinsey.com/insights/sustainability/connecting_climate_change_and_economic_recovery

OECD. (2001). *Decoupling: A conceptual overview [EB/OL]*. OECD. Retrieved from http://www.oecd.org/agriculture/agricultural-policies/25481500.pdf

Popp, D., Newell, R. G., & Jaffe, A. B. (2009). *Energy, the environment, and technological change* (NBER Working Paper No. 14832).

Porter, M. E. (1991). Toward a dynamic theory of strategy. *Strategy Management Journal, 12*(Winter Special Issue), 95–117.

Qingwen, Y. (2012). Measurement of energy efficiency and its relations with environmental policy tools. *Economic Research (Taiwan), 48*(2), 287–317.

Scholz, C. M., & Ziemes, G. (1999). Exhaustible resources, monopolistic competition and endogenous growth. *Environmental and Resource Economics, 13*, 169–185.

Schou, P. (2000). Polluting nonrenewable resources and growth. *Environmental and Resource Economics, 16*, 211–227.

Schou, P. (2002). When environmental policy is superfluous: Growth and polluting resources. *Scandinavian Journal of Economics, 104*, 605–620.
Shiyi, C. (2009). Energy consumption, carbon dioxide emissions and the sustainable development of Chinese industry. *Economic Research, 04*, 41–55.
van Zon, A., & Yetkiner, I. H. (2003). An endogenous growth model with embodied energy-saving technical change. *Resource and Energy Economics, 25*, 81–103.
Wen, X., & Zhaoxi, T. (2011). Energy constraints, technological change and sustainable development—An analytical framework based on the improvement of the quality of intermediate products. *Economic Theories and Management, 01*, 87–94.
Wen, X., & Zhaoxi, T. (2012). Renewable energy, quality of intermediate products and sustainable development. *The World Economy, 02*, 143–159.
Xiaoma, T., Jianwu, X., Xin, H., et al. (2009). A study of energy price distortions and factor substitution in the Chinese industrial sector. *The Journal of Quantitative and Technical Economics, 11*, 3–16.
Yuyan, Z., & Qingyou, G. (2007). World energy landscape and China's energy security. *The World Economy, 09*, 17–30.
Zhou, H., & Tamas, M. M. (2010). Impacts of integration of production of black and green energy. *Energy Economics, 32*(1), 220–226.
Zhou, P., Ang, B. W., & Poh, K. L. (2008). A survey of data envelopment analysis in energy and environmental studies. *European Journal of Operation Research, 189*(1), 1–18.

Chapter 2
Literature Review

Energy-saving technological change has been a hot topic in the energy-economic-environmental (3E) discussions in Western developed countries. As the issues of energy security, climate change mitigation and energy efficiency improvement ascend on government agendas around the world, ESTC has attracted attention from researchers of different disciplines and specialties. While a considerable amount of knowledge about ESTC has accrued, there is a great divergence of views among the researchers, who even have yet to reach a consensus on the basic concept of ESTC. That is why the theory of ESTC is a field that deserves study and has much room for improvement.[1]

2.1 R&D-Based Endogenous Growth Theory

Both the Solow Growth Model and neoclassical growth theory posit that technological change is exogenous, concluding that capital accumulation and innovation are important sources of technological change and economic growth. In contrast to the Solow Growth Model and neoclassical growth theory, endogenous growth theory claims that technological change is endogenously determined by the optimal behavior of economic actors (including consumers, manufacturers and government), and that government policies affect economic growth through influencing the optimal behavior of economic actors. Depending on different mechanisms of endogenous technological change, Romer (1993) categorizes endogenous growth theory into two varieties: capital-based and idea/R&D-based. The former emphasizes the importance of capital accumulation (including physical capital and human capital) in technological change and economic growth, while the latter places weight on R&D

[1] For the sake of brevity, the time subscript is omitted in the equations in this chapter unless otherwise noted.

X. He, *The Endogenous Energy-Saving Technological Change in China's Industrial Sector*, https://doi.org/10.1007/978-981-19-7485-4_2

activities and knowledge accumulation. As the theory of endogenous technological change itself is sprawling and complex, the present paper is chiefly concerned with the mechanisms and channels through which R&D heterogeneity affects ESTC, and this chapter focuses on the basic literature on idea/R&D-based endogenous growth models. A more comprehensive, detailed examination of endogenous growth theory can be found in Pan Shiyuan & Shi Jinchuan (2002), and Yan Chengliang & Gong Liutang (2009). According to Ang & Madsen (2011), R&D-based endogenous growth theory consists of first-generation endogenous growth theory, semi-endogenous growth theory and Schumpeterian growth theory, which are reviewed and critiqued below respectively.[2]

2.1.1 First-Generation Endogenous Growth Theory

The first-generation models of endogenous growth theory are represented by Romer (1990), Segerstrom et al. (1990), Grossman & Helpman (1991), and Aghion & Howitt (1992). All of them assume that technological change is positively correlated with R&D investment and that the knowledge production function has scale effects, which means that new knowledge is proportional to the existing stock of knowledge. Take the knowledge production function proposed by Romer (1990) for example:

$$\frac{\dot{Q}}{Q} = \delta L_Q \tag{2.1}$$

where $\dot{Q}$ is new knowledge, Q is the stock of knowledge, $\frac{\dot{Q}}{Q}$ is the rate of knowledge accumulation; L_Q is the quantity of labor invested in R&D, i.e., R&D investment; and δ is the marginal productivity of the knowledge production function. The scale effect in (2.1) is reflected in a linear relationship between knowledge accumulation and R&D investment, where $\dot{Q}$ will increase with Q, with other factors held constant. Given the difficulty in obtaining new knowledge, the real knowledge production function should be represented as decreasing returns to scale. That is to say, the greater the stock of knowledge is, the harder it becomes to generate new knowledge. The scale effect assumption in (2.1), therefore, is somewhat controversial.

Another key assumption of the first-generation models of endogenous growth theory is that there is no population growth, leading to one more difficulty of explanation. For example, Romer's (1990) steady-state balanced growth path (BGP) is:

---

[2] Yan Chengliang and Gong Liutang (2009) categorize the theory of R&D-based endogenous growth into early Schumpeterian endogenous growth models, Schumpeterian semi-endogenous growth models and Schumpeterian fully-endogenous growth models, which respectively correspond to first-generation endogenous growth theory, semi-endogenous growth theory and Schumpeterian growth theory surveyed in this chapter. Regarding the categorization of R&D-based endogenous growth theory, Yan Chengliang and Gong Liutang (2009) disagree with Ang and Madsen (2011) only on nomenclature.

$$g = g_Q = \delta^* L \tag{2.2}$$

where g denotes the steady-state growth rate; the steady-state growth rate in the absence of population growth equals the steady-state knowledge growth rate g_Q; δ^* is the steady-state R&D share in the amount of labor; and L is the constant amount of exogenous labor. Similar to (2.1), (2.2) also has a scale effect: the greater the amount of exogenous labor L, the higher the steady-state growth rate. The first-generation endogenous growth theorists such as Romer offer the following explanation for (2.2): first, larger L means more R&D investment in an economy, making it easier to produce new knowledge; second, larger L means greater market size, so that innovation enables higher profitability and encourages more R&D investment. For this reason, (2.2) suggested by Romer and other researchers is fairly reasonable.

The two scale effects implied in the first-generation models of endogenous growth theory have caused much controversy. Temple (2003) opines that the scale effect assumption in (2.1) is a knife-edge condition, which heavily constrains the theory. Through empirical studies of OECD countries' economic growth after World War Two, Jones (1995a, b) finds that the scale effect described in (2.2) does not exist in a real-world economy.

2.1.2 Semi-endogenous Growth Theory

With respect to the controversy caused by first-generation endogenous growth theory, especially to the potential defects of the scale effect assumption, the subsequent literature makes two revisions to the theory by assuming that the marginal productivity of the stock of knowledge diminishes over time, and that there are two types of R&D sector within an economy. The first revision has led to what is called semi-endogenous growth theory, represented by Jones (1995b), Kortum (1997), and Segerstrom (1998). They all assume that the marginal productivity of the stock of knowledge in the knowledge production function decreases. For example, Jones (1995b) revises (2.1) as follows:

$$\frac{\dot{Q}}{Q} = \delta L_Q^{\lambda} Q^{\phi - 1} \tag{2.3}$$

where Q, L_Q and δ have the same meanings as in (2.1); $0 < \lambda < 1$ represents an duplication effect, meaning that different productive sectors may make repetitive R&D investment to generate the same knowledge, resulting in lower average productivity of R&D activities; and ϕ represents externalities in knowledge production. Jones (1995b) assumes that ϕ is less than 1, meaning that a larger stock of knowledge renders it more difficult to make new discoveries, a core assumption shared by the semi-endogenous growth theorists. Jones' (1995b) other core assumption is that population growth exceeds zero. Under the above-described assumptions, Jones (1995b) presents a steady-state BGP:

$$g = \lambda n(1 - \phi)^{-1} \tag{2.4}$$

where n is the population growth rate. Equation (2.4) shows that steady-state growth depends on exogenous parameters, including the duplication effect parameter λ, the externality parameter ϕ and the population growth rate n. That is why this variety of model is regarded as "semi-endogenous": while the steady-state growth rate is determined by the optimal behavior of economic actors, government policies do not affect economic growth.

Jones (1999) concludes that firstly, steady-state growth is positively correlated with population growth; secondly, steady-state growth depends on exogenous parameters, which limit government policies' ability to influence steady-state growth. However, neither conclusion is consistent with real-world economic conditions. Kelley (1988) finds that negative population growth rates coexist with positive economic growth rates in Nordic countries; and Kocherlakota & Yi (1997) identifies a stable cointegration relationship between tax policy and economic growth.

2.1.3 Schumpeterian Growth Theory

Schumpeterian growth theory assumes two types of R&D sector present in the economy, in an attempt to cure the defects of the first-generation models of endogenous growth theory. Representatives of the theory include Aghion & Howitt (1998), Dinopoulos & Thompson (1998), Peretto (1998), Young (1998), Jones (1999), Howitt (1999), and Peretto & Smulders (2002). The theory retains the assumption of the first-generation endogenous growth models about the knowledge production function. Unlike the first-generation models, however, Schumpeterian growth theory assumes that horizontal and vertical innovation sectors coexist in the economy to produce knowledge, and with economic expansions, the number of horizontal innovation sectors increases proportionally. Taking the economic system constructed by Jones (1999) as an example, the knowledge production function of the first-generation endogenous growth models are revised as follows:

$$N = \eta L \tag{2.5}$$

$$\frac{\dot{Q}}{Q} = \delta L_{Q_i} \tag{2.6}$$

Equation (2.5) is the knowledge production function of horizontal innovation sectors, while (2.6) represents that of vertical innovation sectors. In the two equations, L is the total labor force, η the marginal productivity of labor in horizontal innovation sectors, and L_{Q_i} R&D investment in No. i vertical innovation sector. Assuming a greater-than-zero population growth rate, Jones (1999) presents the following steady-state BGP under the above setting:

$$g = (\alpha - 1)n + \delta\eta^{-1}s^* \tag{2.7}$$

where α is the marginal output elasticity of horizontal innovation sectors, s^* is the steady-state share of R&D investment in vertical innovation sectors. As seen in (2.7), the rate of population growth n is irrelevant with the steady-state share of labor used in vertical innovation sectors s^*. That means that, even with no population growth, sustainable economic growth is possible. Moreover, government policies may further influence steady-state economic growth by changing s^*.

Schumpeterian growth theory has more or less amended the shortcomings of the first generation and semi-endogenous growth theories and is able to depict the real economy in a better way. But it is no perfect either. As shown in (2.7), it still incorporates a scale effect that links population growth positively with economic growth. In addition, (2.5) and (2.6) may contain too strong assumptions about the knowledge production function. As Li (2000) points out, under the knowledge production functions of the broader two R&D sectors, whether real-world economies fit the conditions described by Schumpeterian growth theory still depends on certain knife-edge conditions.

Despite the large differences in assumptions and settings between the three models of endogenous growth theory, their core idea remains the same: R&D investment is an important source of technological change and economic growth. Ha & Howitt (2007) and Madsen (2008) summarize the parameters of the three endogenous growth models by setting a broader knowledge production function. Equation (2.8) is the knowledge production function built by Ha & Howitt (2007) and Madsen (2008):

$$\frac{\dot{Q}}{Q} = \lambda(\frac{X}{M})^{\varepsilon} Q^{\phi-1} \tag{2.8}$$

where $M \propto L^{\beta}$ in steady-state; X is R&D investment; M denotes varieties of products; ε is the duplication effect; λ is R&D productivity; and ϕ represents externalities in knowledge production. Table 2.1 lists the parameters necessary for the three models of endogenous growth theory.

Table 2.1 Parameters necessary for endogenous growth models based on different types of R&D investment

	ϕ	β	ε
First generation models	=1	=0	>0
Semi-endogenous growth models	<1	=0	>0
Schumpeterian growth models	=1	=1	>0

Sources Ha & Howitt (2007) and Madsen (2008)

2.2 ESTC from the Perspective of Causes

The theory of ESTC originated mainly from two sources: one was the attempt to incorporate energy in economic growth theory, particularly endogenous growth theory, and the other was the economics of climate change. *The Limits to Growth* was the first work to consider systematically the relationship between the constraint of exhaustible resources and sustainable growth. Following *The Limits to Growth*, the earliest studies that systematically introduced energy into economic growth theory included the Dasgupta-Heal-Solow-Stiglitz (DHSS) model of the economics of exhaustible resources (Xiao Wen & Tang Zhaoxi, 2011, 2012) created by Stiglitz (1974a, b), Solow (1974), and Dasgupta & Heal (1974). In the economics of climate change, how to build appropriate policy models incorporating ESTC is also the key to formulating and evaluating energy saving and emissions reduction policies (Löschel & Schymura, 2013). In the early 1990s, ESTC was likened to a "black box" and was generally assumed to be exogenous. Later on, it was considered endogenous, as large quantities of literatures concluded that ESTC was not autonomous as depicted by exogenous ESTC theory, rather the result of the interaction of a set of other factors, including learning, changes in input and output prices, and R&D investment. Much of the theoretical transition from exogenous to endogenous ESTC is consistent with the evolution of economic growth theory. One can even say that economic growth theory has supplied a theoretical foundation and an analytical tool for endogenous ESTC.

2.2.1 Exogenous ESTC

In the 1970s, Western countries underwent a severe oil crisis. In order to explore energy constraints on economic growth, Dasgupta & Heal (1974), Stiglitz (1974a, b), and Solow (1974) built the DHSS model introducing exhaustible resources as a production factor into the neoclassical production function on the basis of the Solow Growth Model to study how the scarcity of resources affect long-term economic growth. Prior to the DHSS model, neither the standard Solow Growth Model nor neoclassical growth theory based on the Ramsey and Diamond models considered the influence of exhaustible resources on economic growth, except Hotelling (1931) and *The Limits to Growth*. Assuming that energy is complementary to other production factors, and there is no technological progress in the economy, *The Limit of Growth* draws a pessimistic conclusion. The DHSS model, on the other hand, believes that so long as there is technological progress in the economy, output and consumption may grow positively in the long run even if the stock of natural resources is limited. Stiglitz's (1974a, b) steady-state BGP is:

$$g = \frac{a_2}{1 - a_1} n + \frac{a_3}{1 - a_1} g_E + g_Q \tag{2.9}$$

where a_1, a_2, a_3 stand for the marginal output elasticity of capital, labor and energy respectively; g_E is the growth rate of energy consumption; and g_Q is the rate of ESTC. Stiglitz (1974a, b) assumes a negative growth rate of long-term energy consumption (the extraction rate of energy) but positive constant rates of population growth and technological change. As shown in (2.9), as long as the rate of technological change exceeds a certain positive threshold, greater-than-zero long-term growth rates of per capita output and consumption can be guaranteed even with limited energy supplies. In fact, as technological change in the DHSS model is generally assumed to be exogenous and Hicks-neutral, strictly speaking, ESTC in the DHSS model is incapable of measuring energy efficiency comprehensively. Because capital, labor and energy produce the same rate of technological change under the assumption of Hicks-neutral technological change, one can hardly discern whether observed improvements in energy intensity are owing to the crowding out of energy input by other production factors or to any increase in effective energy input itself.

After the DHSS model, Dasgupta & Heal (1980) studied the issue of sustainable growth under the constraint of exhaustible resources when the substitution elasticities of natural resources and other production factors are less than 1. Dasgupta & Heal (1980) find that even if natural resources are complementary to other production factors, the doomsday scenario predicted in *The Limits to Growth* will not arrive, as long as positive resource-augmenting technological change is present in the economy.

In climate change economics, autonomous energy efficiency improvement (AEEI) is often used as an exogenous parameter of the decoupling between economic growth and energy consumption. AEEI is usually assumed to be a function of time, and is mainly based on the assumption of exogenous technological change in neoclassical growth theory, in a way similar to how technological change is treated in the DHSS model. Unlike the DHSS model, however, AEEI assumes that technological change in a productive sector tends to be achieved through increased effective energy input. According to Löschel & Schymura (2013), the simplest AEEI can be expressed as:

$$Y = F(K, L, Q_E E) \tag{2.10}$$

where Y is output; K is capital stock; L is labor input; E is energy input; $Q_E = e^{rt}$ represents AEEI; and r is an exogenous parameter. Early application of AEEI in the study of sustainable growth was found in the book of Nordhaus (1994) and McCracken et al. (1999), both of which introduce AEEI into the production function, and evaluate the effectiveness of energy and emissions reduction policies in the context of different exogenous ESTC rates. Recent studies assume that AEEI takes more complex forms, and introduce it into economic models of climate change. For example, Jacoby et al. (2006) assume that the form of AEEI varies across different regions; Richels & Blanford (2008) divide AEEI into pessimistic, moderate and optimistic categories based on different r values, and investigate U.S. emissions reduction potentials at various rates of ESTC.

AEEI has the advantages of a simple, transparent form and ease of sensitivity analysis (Popp et al., 2009). In contrast to the DHSS model, (2.10) distinctively differentiates the substitution effect leading to lower energy intensity from the effect of increased effective energy input, assuming that effective energy input increases with time. However, the shortcomings of AEEI are also obvious. Through their empirical studies, Oravets & Dowlatabadi (1996) find it hard to fix the form of AEEI in actual modeling. According to Löschel (2002), by regarding technological change as a "black box", AEEI overlooks the possible impact of energy price changes and R&D investment on ESTC. Sue Wing (2006) holds that AEEI only allows incremental technological changes in the model without regard to the potential impact of major technological breakthroughs on ESTC. Löschel & Otto (2009) point out that AEEI turns a blind eye to the possible uncertainties in technological change.

On the whole, the DHSS model and AEEI, by assuming exogenous ESTC, expound the logic that technological change can effectively overcome energy constraints and achieve sustainable economic growth. Compared with the DHSS model, AEEI distinguishes between the effect of factor substitution and the effect of increased effective energy input. As to how technological change occurs, however, neither the DHSS model nor AEEI gives a clear clue.

2.2.2 Semi-endogenous ESTC

Semi-endogenous ESTC is mainly concerned with the backstop technology assumption in climate change economics, an early attempt by climate change economists to make ESTC endogenous after AEEI. Backstop technologies refer to those technologies that have been known but have not yet been commercialized on a large scale, such as nuclear, wind and solar power in the early years of their development. With backstop technologies maturing and their costs decreasing, the existing technologies will become relatively expensive and eventually be replaced by backstop technologies (Löschel, 2002). There are two approaches to modeling a backstop technology in the extant literature (Löschel & Schymura, 2013). The first is to assume that the supply of a backstop technology is limitless above a fixed marginal cost. The second is to assume that the marginal cost of a backstop technology falls at a fixed rate. The first approach is similar to the "structural break" (Löschel & Schymura, 2013), while the second idea is akin to AEEI (though it is different from AEEI in that Q_E in a backstop technology acts on the marginal cost of technology, and decreases at a constant rate over time).

The controversy about the backstop technology assumption centers on the fact that it implies a key assumption that existing and backstop technologies are fully mutually substitutable, brushing aside the correlation between advancements in different technologies. Popp (2005) revises the backstop technology assumption by limiting the penetration rate of backstop technologies to allow an incomplete substitution relationship between existing and backstop technologies. Because the marginal cost of a backstop technology in (2.10) in the literature c is exogenous,[3] and p_E is determined by the optimal behavior of economic actors in the model, backstop technologies are also called "semi-endogenous ESTC" (Sue Wing, 2006). Since actually backstop technologies are not entirely determined by the optimal behavior of economic actors, that is, they are not fully endogenous, the backstop technology assumption has the same flaws as the DHSS model and AEEI.

2.2.3 Endogenous ESTC

As exogenous growth models give way to endogenous ones, leading edge research on ESTC has attempted to introduce natural resources into endogenous growth models. The research is focused on R&D-based endogenous growth theory, but the current literature reflects two different veins of theory-building logic: one is to introduce natural resources into the product-diversification model built by Romer (1990), the other is to incorporate natural resources in the quality-ladder framework developed by Grossman & Helpman (1991), and Aghion & Howitt (1992). Meanwhile, climate change theorists have tried to make ESTC endogenous by suggesting three ideas: price-induced technological change (PITC), originated by Hicks (1932); learning-induced technological change (LITC), originated by Arrow (1962); and R&D-induced technological change (RITC), originated by the first-generation endogenous growth theorists (Kamien & Schwartz, 1968; Binswanger & Ruttan, 1978; Romer, 1990; Grossman & Helpman, 1994). However, the theory of R&D-induced technological change in climate change economics is somehow different from the theory-building logic that directly introduces energy into endogenous growth models to deal with ESTC. This chapter reviews and critiques the major literatures on PITC, LITC, and RITC. As in the relevant literature most researchers introduce energy into R&D-based endogenous growth theory, in the interest of simplicity, I use the term RITC to cover R&D-induced technological change both in the R&D-based endogenous growth literature and in climate change economics in the following sections, and compare the similarities and differences between endogenous growth theory and R&D-based endogenous ESTC in climate change economics.

(1) Theory of price-induced technological change

The theory of price-induced technological change originated from a hypothesis suggested by Hicks (1932) that changes in the relative prices of production factors will induce people to use expensive factors in a more economical manner, or use other cheaper substitutable factors. In the wake of the 1970s oil crisis, studies of ESTC emerged and prevailed in Western academic communities. A large body of literature has highlighted the question of whether higher energy prices will induce ESTC,

[3] In the recent literature, it is assumed that the marginal cost of a backstop technology decreases at an exogenous rate, or there is a learning-by-doing effect in backstop technologies. The learning rate in learning-by-doing, however, is still exogenous (Popp et al., 2009).

leading to less use of fossil fuels. Kennedy (1964) was among the first to translate Hicks' hypothesis into a theoretical system. Ahmad's (1966) empirical studies find that hikes in relative prices of fossil fuels often result in greater energy efficiency and pervasive use of other substitute energy sources. The core of PITC theory is that higher energy prices will bring about ESTC, leading to energy efficiency improvements. In climate change economics, PITC is generally assumed to be an efficiency parameter associated with energy prices.

There have been few studies that apply PITC alone in economic analysis in recent years. Researchers engaged in typical PITC-related research include Popp (2002), who empirically finds that rising energy prices can significantly boost the number of energy-related patents; Abadie & Chamorro (2008), who discover that implementation of a carbon tax may quicken the pace of ESTC; Dowlatabadi (1998), who combines AEEI with PITC to initiate a price-induced efficiency theory, and on that basis estimates the benefits and costs of different energy and emissions reduction policies; and Jakeman et al. (2004), who apply PITC in their model and compare the costs of implementing the Kyoto Protocol with and without ESTC, respectively.

Extant empirical studies show that PITC only partially explains ESTC, though higher energy prices do stimulate ESTC (Newell et al., 1999; Popp, 2002). For this reason, there are fewer and fewer research projects that use PITC alone to construct and assess energy-saving and emissions reduction policies.

(2) Theory of learning-induced technological change

The idea of LITC originated from Arrow (1962), who holds that the unit cost of production declines with the accumulation of production experience (e.g., output, installed capacity, R&D investment, etc.), and refers to the effect as learning-by-doing. Central to the idea is the learning rate, which measures how fast the average cost falls with the accumulation of experience. So far, no real consensus has emerged on the methodology for empirical studies of learning-by-doing (Thompson, 2012). In the extant literature, (2.11) is often cited to represent the basic form of learning-by-doing (Löschel & Schymura, 2013):

$$ct = y^{-b} \tag{2.11}$$

where ct is the unit cost of producing a certain product; y is the output of that product; $b > 0$ denotes the learning rate. In (2.11), the unit cost of production is in inverse proportion to the output. From (2.11), Löschel (2002) goes one step further to define 2^{-b} as the progress rate measuring how much unit cost will decrease when the output doubles. The more recent literature expands Arrow's (1962) definition of "experience", and by the source of "experience" breaks down learning-by-doing into learning-by-using (Rosenberg, 1982), learning-by-researching (Cohen & Levinthal, 1989), and a Two-Factor Learning Curve that combines learning-by-doing and learning-by-researching (Kouvaritakis et al., 2000a, b). Since LITC emphasizes the learning of technology, it is generally used to depict ESTC in bottom-up models,[4] such as Messner & Schrattenholzer's (2000) MESSAGE model, Capros & Mantzos's (2000) PRIMES model, Kouvaritakis et al.'s (2000a, b) POLES model, and Loulou et al.'s (2005) TIMES model.

Critics of LITC theory declare that the LITC assumption is ad-hoc and lacks transparency in modeling, making it extremely difficult to sum up the learning rates in different sectors, especially when the rates of experience accumulation vary across different sectors or different economic actors within the same sector. In addition, the difficulty in identifying the trigger behind learning-by-doing is the major challenge the theory of learning-by-doing has to face, as correct identification of that trigger is the foundation of (2.11) (Popp et al., 2009). Therefore, research on RITC theory and its application has received increasing attention.

(3) Theory of R&D-induced technological change

(1) R&D-induced ESTC in endogenous growth theory

Current research on R&D-induced ESTC in endogenous growth theory is mainly based on the first-generation models of endogenous growth theory, and evolves in two directions depending on whether technological change is achieved through horizontal or vertical innovation. Scholz & Ziemes (1999), Schou (2002), and van Zon & Yetkina (2003) draw on Romer's (1990) approach to study endogenous ESTC from the perspective of product diversification, namely, horizontal innovation. Aghion & Howitt (1998, 2011), Grimaud & Rouge (2003), and Xiao Wen & Tang Zhaoxi (2011, 2012), on the other hand, employ the frameworks of Grossman & Helpman (1991), and Aghion & Howitt (1992), respectively, to define endogenous ESTC in terms of quality improvement, namely, vertical innovation. As technological change in Grossman & Helpman (1991) and Aghion & Howitt (1992) is reflected in the improvement of capital quality, the model built through vertical innovation can be called an embodied technological change model. In Xiao Wen & Tang Zhaoxi (2011), for example, the production function of finished goods manufacturing sectors is:

$$Y = H_Y^{a_1} \int_0^1 Q(i)x(i)^{a_2} di E^{a_3} \tag{2.12}$$

where $Q(i)$, $x(i)$ and E represent the quality of intermediate goods, intermediate goods and energy respectively; and production factors satisfy the condition of constant returns on scale, i.e., $a_1 + a_2 + a_3 = 1$. As shown in (2.12), under the

[4] Bottom-up models, such as the MARKAL model, start from engineering technology, and optimize the energy system by elaborating on energy production and consumption technology. In contrast to them are top-down models, which describe the performance of the economy, including production, consumption, and import and export, but seldom address technology employed in the energy system. For discussions of bottom-up and top-down models in more detail, read Löschel and Schymura (2013). In addition, there have appeared hybrid models in recent years, which blend the two modeling methods. For a comparison of modeling methods, see Sue Wing (2006).

Cobb–Douglas production function, improvement of the quality of intermediate goods stimulates intermediate goods to replace labor and energy, and in turn improves energy efficiency. However, Xiao Wen & Tang Zhaoxi (2011) basically disregard the possibility that more effective energy input could improve energy efficiency.

Hassler et al. (2012), based on the theory of directed technical change (Acemoglu, 2002), developed a model of R&D heterogeneity-based endogenous ESTC, which partially solved the problems in (2.12). According to Hassler et al. (2012), capital-labor and energy produce different factor-augmenting technological changes, which come from different R&D sectors. They set a production function for finished goods manufacturing sectors as follows:

$$Y = \min\{QK^{a}L^{1-a}, Q_E E\} \tag{2.13}$$

where Q and Q_E represent capital-labor-augmenting and energy-augmenting technological change respectively. Because Q and Q_E simultaneously enter the production function, Hassler et al.'s (2012) model is able to measure energy efficiency by whether effective energy input is increased. But the limitations of (2.13) are evident. Firstly, it claims that capital-labor and energy are fully complementary to each other (Hassler et al. (2012) refer to that as the "Leontief condition"). As the supply of energy tends to dwindle over time, Q_E must be greater than a certain positive threshold to sustain positive economic growth. Because of the fully complementary relationship between capital-labor and energy, (2.13) contains an implied assumption that as capital and labor inputs increase, Q and Q_E may move reversely near BGP, that is, energy-augmenting technological change accelerates whereas capital-labor-augmenting technological change slows down, though no empirical studies have lent support to this assumption yet.[5] Secondly, while (2.13) assumes that capital-labor-augmenting technological change and energy-augmenting technological change enter the production function both in the form of disembodied technological change, existing empirical studies have found the presence of significant embodied technological change in the economy, especially capital-augmenting technological change in the form of capital quality (Zhao Zhiyun et al., 2007; Song Donglin et al., 2011).

Despite the above-mentioned problems, I applaud Hassler et al. (2012) for their fairly successful, exemplary endogenous ESTC model and their revision of endogenous embodied ESTC, particularly their idea of distinguishing between energy R&D investment and non-energy R&D investment. Their work has provided food for thought for answering the following questions: What are the differences between the effect of factor substitution and the effect of increased effective energy input on

[5] The analytical paradigm of neoclassical economics excludes the possibility that Q is negative and Q_E is positive in (2.13). As indicated in the existing empirical studies, however, Werf (2008), and Jin and Jorgenson (2010) independently find positive Q_E and negative Q in major industrial sectors in 12 OECD countries and the United States. In addition, (2.13) assumes that capital and labor produce common factor-augmenting technological change, but existing empirical research does not fully support this assumption (Werf, 2008; Jin and Jorgenson, 2010).

energy efficiency? Is it practical to integrate these two effects in one framework of analysis? It stands to reason, therefore, that by considering both effects on energy efficiency, one can measure energy efficiency more comprehensively.

(2) R&D-induced ESTC in climate change economics

R&D-induced ESTC in climate change economics is mainly inspired by Kamien & Schwartz (1968), Binswanger & Ruttan (1978), and the first-generation endogenous growth theorists (especially Romer (1990), and Grossman & Helpman (1994)). The R&D market is not an integrated one because R&D outcomes in one sector will spill over to others, and increases in energy R&D investment are likely to crowd out other types of R&D investment (Popp et al., 2009). As demonstrated by empirical studies, the crux of the current theory of R&D-induced technological change is how to balance the spillover and the crowding-out effect of R&D investment, because the spillover effect often brings about greater ESTC, while the crowding-out effect does otherwise (Löschel, 2002; Popp et al., 2009; Löschel & Schymura, 2013).

In recent years, the theory of R&D-induced technological change has become the mainstream methodology for endogenous technological change in climate change economics, and a series of classic literatures have emerged. For example, Goulder & Schneider (1999) introduce R&D-induced technological change into a climate policy model, and, based on an analysis of the knowledge spillovers among four types of productive sector, find that achieving stated emissions reduction targets incurs only a little cost when there is endogenous technological change. Goulder & Mathai (2000), assuming that knowledge accumulation would lower emissions reduction costs, have developed a learning curve for energy and emissions reduction sectors. Nordhaus (1992) has built the R&DICE model by increasing total knowledge stock in the DICE model. He finds that any increase in knowledge has a limited impact on carbon dioxide intensity if no subsidy is provided for R&D activities, with differences in R&D returns between society and individuals as an indicator of technology spillovers. Popp (2004) has developed his ENTICE model based on Nordhaus (2002), introducing knowledge as an alternative input into the energy production function, and finding R&D expenditure will crowd out other expenditures. On the basis of Goulder & Schneider (1999), Sue Wing (2001, 2003) has set up a Dynamic Computable General Equilibrium (DCGE) model incorporating 89 productive sectors and simulated the effects of different R&D subsidy levels and emissions reduction targets on the U.S. economy. Sue Wing (2001, 2003) differs from Goulder & Schneider (1999) in that the former considers the relative price changes and their general equilibrium consequences probably caused by different R&D subsidy levels. Buonanno et al. (2003), assuming in their ETC-RICE model that the total stock of knowledge worldwide can enter the production function of each country, finds that under a strong R&D spillover setting, knowledge accumulation can effectively reduce carbon dioxide intensity. The DEMETER-2 model developed by Gerlagh & Lise (2005) includes an R&D sector that can produce knowledge, which will enter the production function for various energy sources as alternative inputs, leading to "learning-by-researching" in productive sectors. Otto et al. (2008) find, by simulating based on Acemoglu (2002) the relationships among energy, technological change and economic growth under different emissions restriction policies, that factor substitution allows technological change to help save energy. Using a model similar to Sue Wing's (2001, 2003), Wang et al. (2009) have built a Chinese DCGE model involving 41 productive sectors. Through analyzing the marginal costs of different R&D subsidy levels and emissions reduction targets, Wang et al. (2009) find that in the absence of R&D subsidy policies, technological change will not directly lead to absolute reductions in carbon emissions. Bosetti et al. (2009), Bosetti & Tavoni (2009), and Bosetti et al. (2011) discover, by analyzing the impact of different innovation policies on carbon emissions in the WITCH model, that knowledge accumulation can significantly reduce carbon emissions.

As informed by the existing literature, the gist of R&D-induced ESTC theory is that knowledge accumulation will crowd out energy input. The theory of R&D-induced technological change developed by Sue Wing (2001, 2003) and Wang et al. (2009) is shown as an example in (2.14):

$$Y = F(H, K, L, E, M) \equiv \left[\gamma_H H^{\frac{\sigma_H - 1}{\sigma_H}} + (1 - \gamma_H)(KLEM)^{\frac{\sigma_H - 1}{\sigma_H}}\right]^{\frac{\sigma_H}{\sigma_H - 1}} \tag{2.14}$$

where Y is total output; H is the stock of knowledge; *KLEM* represents a composite of capital, labor, energy and intermediate inputs; σ_H is the elasticity of substitution between knowledge and the composite; and γ_H is a share parameter. Although the setting of (2.14) is widely used in climate change economics, the chief reason that R&D-induced technological change improves energy efficiency, as shown by (2.6), is that knowledge accumulation replaces energy as used in past production, especially when the elasticity of substitution between knowledge and the composite is greater than or equal to 1. As can be seen, the theory of R&D-induced technological change in climate change economics also ignores the effect of increased effective energy input on energy efficiency.

2.3 ESTC Seen Through the Lenses of Different Research Methods

This section examines the empirical research methods and outcomes relating to ESTC recorded in the existing domestic and foreign literatures.

2.3.1 Econometric Methods

(1) Current State of Research in Foreign Countries

(1) Macro-econometric Method

The macro-econometric method for ESTC, like that applied in macroeconomics, is based on long-term time series data, which provides a sound empirical foundation. Currently, empirical research on ESTC by the macro-econometric method often follows the traditional method of analyzing macro-econometric models, and building macro-econometric models pursuant to New Keynesian Economics. For this reason, this method is especially appropriate for short-term and medium-term forecasting (Carraro, 2002).

Jorgenson & Wilcoxen (1993) have constructed a granular U.S. energy economic model to compare the economic impacts of different tax policies in achieving the same emissions reduction goal. They find that the carbon tax has the least impact on the economy in view of the set emissions reduction targets. The WARM model developed by Carraro & Galeotti (1997) shows that energy R&D investment, technological innovation and diffusion can offer right incentives to enterprises, and help them stay competitive in the marketplace without damaging the environment. The E3ME model introduced by Barker & Köhler (1998) considers the distribution effect of the energy tax, concluding that energy tax refunds to consumers in the form of a lump-sum tax can avoid the distortion caused by the tax and maximize the well-being of society at large. Von Zon & Yetkiner (2003) suggest that converting the energy tax to an R&D subsidy to encourage productive sectors to invest in energy and emissions reduction efforts will accelerate R&D-induced technological change and stimulate economic growth. In his model, Gerlagh (2008) examines R&D investment for both clean energy and fossil fuels, and observes that as fossil fuel prices rise and energy and emissions reduction policies are implemented, returns from clean energy R&D will exceed those from fossil fuel R&D over time, making it possible for clean energy R&D to crowd out fossil fuel R&D.

(2) Micro-econometric Method

Empirical micro-econometric analysis of ESTC is based on the characteristics of different industry sectors or products. Pakes et al. (1993), analyzing the relationship between the fuel economy of vehicles on the market and gasoline prices in the U.S., find that increased energy prices after the oil crisis have driven substantial improvements in fuel economy. Berry et al. (1996) find that quality-adjusted costs in the U.S. automotive industry have increased significantly with higher fuel costs and emissions standards. Newell et al. (1999) empirical analysis of whether increases in energy prices will affect the energy efficiency of new durable consumer goods reveals that energy-saving technological innovations such as room air conditioners and gas water heaters have been driven by energy price changes. Using U.S. patent data from 1970 to 1994 to analyze the impact of energy prices on energy efficiency innovations, Popp (2002) detects a strong positive correlation between energy prices and energy

R&D. Based on his analysis of the relationship between energy price hikes and the adoption of energy-saving technology in U.S. manufacturing sectors, Linn (2008) finds that a ten percent increase in the price of energy leads to one percent rise in the adoption of energy-saving technology. Sue Wing (2008) surveys the number of patents at the industry level in the U.S. to conclude that R&D induced technological change driven by energy prices is chiefly responsible for declines in energy intensity in the U.S.

(2) Current State of Research in China

Li Shixiang & Cheng Jinhua (2008), analyzing the factors affecting energy efficiency in China by a two-step method based on data collected in 29 provinces from 1990 to 2006, find that industrial restructuring and technology upgrading have played a central role in improving energy efficiency. Zhou Jian (2008), using panel data of 31 Chinese provinces from 1989 to 2005 and spatial econometric techniques, finds that the energy efficiency of each province shows meaningful spatial dependence. Wu Yuming & Jia Lin (2009) identify a random convergence of energy efficiency in different areas, based on data of 30 Chinese provinces from 1997 to 2007. Zhang Shaojun & Li Dongfang (2009) empirically study the relationship between non-integrated production and energy efficiency using panel data on 33 industrial sectors of China from 1992 to 2005. Through an empirical analysis of the rebound effect using the Translog model, Guo Juan et al. (2010) recognize a general trend of improving energy efficiency in China's industrial sectors. Shao Shuai et al. (2010) uncover an inverted N-shaped relationship between energy intensity and labor productivity after analyzing based on the STIRPAT model the factors affecting energy intensity and carbon emissions of Shanghai's industrial sectors from 1994 to 2008. Wang Keliang et al. (2010) estimate the energy efficiency and environmental performance of different provinces based on provincial panel data from 2000 to 2007, and empirically analyze the factors affecting the energy saving and emissions reduction potentials of the provinces involved. With panel data on 37 industrial sectors of different provinces from 2000 to 2006, Wang Qiubin (2010) investigates the implications of industrial and regional structures to energy efficiency, finding that industrial structure's role is negative while that of the international division of labor is positive. Yang Zhongdong (2010) focuses on the business cycles and industrialization of the heavy chemical industry and finds, based on data on 27 industrial sectors from 2002 to 2007, that rapid economic expansion and accelerated industrialization of the heavy chemical industry were the major contributors to the declines in energy efficiency in China's manufacturing sectors from 2002 to 2007. Zha Jianping (2011) looks at the relationship among industrial growth, energy consumption and carbon emissions in China from the perspective of "decoupling" and "re-coupling" theory, and believes that from 2000 to 2009 energy efficiency gains in China's industrial sectors mainly drove the weak decoupling between industrial growth and carbon emissions. Chen Demin & Zhang Rui (2012) use panel data of 29 Chinese provinces from 2000 to 2010 to survey how environmental regulation affects energy efficiency. They find that the reaction of energy efficiency to environmental regulation varies greatly from province to province. Based on provincial panel data from 1995 to 2010,

Meng Chang & Chen Yujie (2012) study the volatility of China's energy efficiency, and find that lower resource prices lead to falling energy efficiency. Yuan Yijun et al. (2012) analyze the influence of structural adjustment, technological progress and management improvement on energy efficiency based on panel data of 29 provinces from 2000 to 2010.

2.3.2 *Index Decomposition Analysis*

(1) Current State of Research in Foreign Countries

Index Decomposition Analysis (IDA) disentangles different factors affecting an object of study based on different indexes constructed. Myers & Nakamura (1978) introduced IDA into the area of energy for the first time, and separated energy efficiency from other factors unrelated to energy efficiency. Application of IDA is therefore focused on those factors bearing on energy efficiency and ESTC. Recent developments in IDA research include refining objects of study, such as decomposing them at the industry level (Ang & Zhang, 2000), and explaining the decline of economy-wide energy efficiency with bottom-up models (Ang, 2006). Depending on decomposition methods, IDA comes in two types: Laspeyres Decomposition Method (LDM)-based and Divisa Decomposition Method (DDM)-based. Both LDM- and DDM-based methods allow additive decomposition and multiplicative decomposition. Common decomposition methods include the LDM-based Fisher Ideal Index method, the DDM-based Log Mean Divisa Index (LMDI) method, and the Arithmetic Mean Divisa Index (AMDI) method.

According to Ang (2004), that the residual term is not negligible in LDM leads to unreliable decomposition results; on the other hand, the LMDI method solves the residual term problem because the factors are reversible, resulting in more reliable conclusions. In the early studies of energy efficiency decomposition, the AMDI method was used by researchers like Wilson et al. (1994), Cox et al. (1997), Harris & Thorpe (2000), and Tedesco & Thorpe (2003). In recent years, researchers, including Sandu & Syed (2008), Sandu & Petchey (2009), and Petchey (2010) have turned to the LMDI method. Sandu & Syed (2008) use overall data in studying the factors affecting energy efficiency in Australia, and subsequently Sandu & Petchey (2009), and Petchey (2010), based on the work of Sandu & Syed (2008), apply more microscopic sample data to their studies on subdivided productive sectors.

(2) Current State of Research in China

Shi Bo (2007) employs LDM to decompose the changes in energy intensity in China from 1980 to 2005 and finds that decreased energy intensity in the industrial sectors was the major cause of energy efficiency improvements. Qin Fangming & Shi Bo (2010) use LDM to decompose the changing trends of coal and oil intensity in China from 1980 to 2007 and find that both coal and oil utilization efficiency led to falls in energy intensity. Wu Qiaosheng (2010) employs the Fisher Ideal Index to decompose the changes in energy intensity from 1981 to 2007 at the industry level, concluding that declines in energy intensity during the period were mainly driven by improvements in energy efficiency rather than structural factors. Guo Xiaoming & Zhang Zongyi (2012) use LMD and LMDI respectively to analyze the factors influencing energy intensity in China's transport industry from 1985 to 2008, finding that efficiency factors explained the majority of the fluctuations in transport energy intensity. Song Feng & Wang Lili (2012) use the Fisher Ideal Index to decompose the energy intensity and its changes in 28 provinces, municipalities or autonomous regions of China from 1995 to 2009, and ascribe falling energy intensities to higher energy efficiency during that period.

2.3.3 Data Envelopment Analysis

(1) Current State of Research in Foreign Countries

Data Envelopment Analysis (DEA) is a non-parametric method of measuring the efficiency of multiple decision making units (DMU) in operational research and frontier analysis in economics. It was the mainstream tool for analysis of the efficiency of the electric power sector in the 1980s (Zhou et al., 2008). In DEA, technological change is defined in terms of engineering (IPCC, 2000). Färe et al. (2011) have extended the method to incorporate endogenous technological change.

With DEA, Chauhan et al. (2006) evaluate the energy efficiency of rice production in India, and find that more efficient tillering and better machinery would effectively improve energy efficiency, saving at least 11.6% of energy use. After investigating the energy efficiency of 32 five-star hotels in Antalya, Turkey, Onut & Soner (2006) conclude that better energy management could effectively make the hotels more profitable. Hu & Wang's (2006) empirical analysis of the energy efficiency of 29 administrative regions in China from 1995 to 2002 reveals a U-shaped relationship between per capita income and total-factor energy efficiency (TFEE). Hu & Kao (2007) calculate the energy-saving target ratios (ESTR) of 17 APEC economies from 1991 to 2000 under a total-factor framework. Bian & Yang (2010) extend the classical Shannon-DEA method to evaluate the energy and environmental efficiency of 30 Chinese provinces. Mukherjee (2008) empirically analyzes the energy efficiency of India's manufacturing sectors with the directional distance function, finding that by improving technical efficiency, the manufacturing sectors can use 3.84% less energy on average every year. Houshyar et al. (2011) study the energy efficiency of raw corn production in Fars, southwestern Iran, and find that changing the existing energy utilization mode can elevate farmers' productivity and in turn increase their income.

(2) Current State of Research in China

Based on their assessment of how energy efficiency changed and was affected in 13 Chinese industrial provinces or autonomous regions from 1990 to 2006, Li Shixiang & Cheng Jinhua (2008) find that energy efficiency varies substantially across the provinces, and that the current relative pricing mechanism fails to provide effective incentives for energy conservation and emissions reduction. Shi Bo & Shen Kunrong (2008) examine TFEE in different Chinese provinces and regions from the perspective of market segmentation, with knowledge incorporated into the production function. They find that TFEE was high and flat in the eastern region, but spiraled in the central and western regions in the 1995–2005 period. Wei Chu & Shen Manhong (2008) have developed an index based on 1995–2006 data to measure TFEE in 29 Chinese provinces. Based on time series data, Zeng Sheng & Huang Dengshi (2009) investigate energy efficiency's relationship with economic growth in the 1994–2007 period, confirming that greater energy efficiency helps boost economic growth. Li Guozhang & Huo Zongjie (2009) research TFEE in Chinese provinces during the 1995–2006 period, finding that TFEE edges higher from west to east and from north to south geographically, and meanwhile nationwide TFEE converges with local TFEE. Pang Ruizhi et al.'s (2009) research on TFEE in 35 industrial sectors of China in the 1994–2007 period shows that both TFEE and technical efficiency were low in these sectors. Based on her examination of TFEE in 30 provinces in the 1990–2006 period, Qu Xiao'e (2009) believes that there were large gaps in TFEE among eastern, central and western regions. Tang Ling & Yang Zhenglin (2009) study energy efficiency in China's industrial sectors from 1998 to 2007, and explore the mechanism whereby industrial transition helped raise energy efficiency, with the conclusion that industrial TFEE changed along a U-shaped curve as enterprises grew larger. Yuan Xiaoling et al. (2009), based on data of 28 provinces from 1995 to 2006, inspect nationwide and local TFEE, and empirically analyze how nationwide TFEE and local TFEE converged and what affected their performance, observing that TFEE was negatively correlated with industrial structure, property right structure, energy consumption structure and resource endowment, but positively correlated with energy prices. Wang Haining & Chen Yuanyuan (2010) apply 2001–2007 data on 25 industrial sectors to their empirical analysis of the relationship between externalities and energy efficiency, concluding that TFEE was low in energy-intensive sectors. Wang Xiao & Qu Xiao'e (2010) appraise energy efficiency in 28 manufacturing sectors in the 2001–2007 period, and find that their TFEE was trending higher, but varied significantly from one sector to another. According to their empirical findings, the differences in energy efficiency can be explained by different levels of technological progress; and R&D, market competition and FDI helped improve energy efficiency in manufacturing. Wei Chu et al. (2010) evaluate how energy efficiency evolved in 30 provinces from 2005 to 2007, and comment on the energy and emissions reduction potentials in different regions.

2.3.4 *Computable General Equilibrium*

(1) Current State of Research in Foreign Countries

Computable general equilibrium (CGE) is a widely used tool for policy simulation and assessment. Well-known CGE models in climate policy research include the GREEN model (Burniaux et al., 1992), the G-CUBEC model (McKibbin & Wilcoxen, 1992), the PACE model (Böhringer et al., 2002), Goulder and Schneider's model (1999), the SGM model (McCracken et al., 1999), the Sue Wing-EPPA model (Sue Wing, 2001), the MIT-EPPA model (Jacoby et al., 2006), and models respectively developed by Smulders & Nooij (2003), Jakeman et al. (2004), Sue Wing (2006), and Wang et al. (2009). The GREEN model is a multi-region and multi-sector DCGE model developed by OECD, which assumes ESTC in AEEI form. The G-CUBEC model is a multi-region and multi-sector inter-temporal computable general equilibrium (ICGE) model, which also assumes ESTC in AEEI form. The PACE model is a multi-sector dynamic CGE model, which adopts bottom-up modeling for energy sectors, and top-down modeling for other sectors, and assumes ESTC in AEEI form. The SGM model is a multi-sector regional dynamic CGE model assuming ESTC in AEEI form. The Sue Wing-EPPA model is a multi-sector dynamic CGE model, which assumes that ESTC is R&D-induced, and simulates the impact of different R&D subsidy levels and emissions reduction targets on the U.S. economy. The MIT-EPPA model is a multi-region and multi-sector dynamic CGE model developed by MIT, which assumes ESTC in a mixed AEEI, price-induced, and learning-induced form. Smulders & Nooij (2003), through introducing the theory of R&D-induced technological change, find that energy R&D investment can effectively reduce the cost of energy saving and emissions reduction policies. Assuming that technological change is price-induced, Jakeman et al. (2004) observe that higher energy prices may help lower emissions reduction costs as they accelerate technological change. Sue Wing's (2006) model is a multi-sector dynamic CGE model. Unlike the Sue Wing-EPPA model, Sue Wing (2006) chooses a hybrid model, assuming that technological change is R&D-induced. Wang et al. (2009) construct a dynamic CGE model incorporating 41 productive sectors based on the Sue Wing-EPPA model. After analyzing the marginal costs of different R&D subsidy levels and emissions reduction targets, they find that without R&D subsidy policy, technological change alone would not have directly led to absolute carbon emissions reductions.

(2) Current State of Research in China

Using a CGE model encompassing various energy sources, Zha Donglan & Zhou Dequn (2010) have identified a significant energy rebound effect in the Chinese economy. Li Aijun (2010) finds in a DCGE model featuring 22 productive sectors that only through substantial improvements in energy efficiency is it possible to boost long-term growth while reducing energy intensity. From the perspective of ensuring economic growth, Yao Xin & Liu Xiying (2010) have designed a dynamic path to the optimal carbon tax for China through a dynamic optimal carbon tax model to maximize the well-being of society under growth constraints. Li Yuanlong & Lu

Wencong (2011) have developed a CGE model to simulate the energy rebound effect in the Chinese economy after energy efficiency improves in productive sectors. They find that for every five percent increase in energy efficiency, the energy rebound effect would fetch 52.38% in the short run and 178.61% in the long run. Through simulating the consequences of reforming fossil fuel subsidies in a CGE model, Yao Xin et al. (2011) find that converting fossil fuel subsidies into energy R&D investment in clean energy sectors would result in less energy use and emissions while sustaining economic growth. Yuan Pengfei & Wu Jilin (2011) simulate in a CGE model the dynamic characteristics of China's energy consumption and economic growth in the context of higher energy prices, and find that higher energy prices would cause significant reductions in GDP, exports and employment, but increases in imports. They conclude that advancing reform of energy market pricing would boost energy efficiency and help attain energy and emissions reduction goals. With a dynamic CGE model featuring R&D-induced technological change, Wei Wei & He Xubo (2012) simulate the time paths of energy-economy-environment variables under different R&D subsidy scenarios. According to them, while R&D subsidy can effectively promote ESTC, excessive subsidy would jeopardize economic growth. Wei Wei & He Xubo's (2013) further simulation of the macro and the sectoral general equilibrium effects of carbon emissions limits and R&D subsidy policy reveals that the two effects work together to mitigate GDP losses arising from implementing emissions limits alone, and help economic actors reduce energy consumption and emissions.

2.3.5 Integrated Assessment Models

Integrated assessment models (IAMs) are designed to assess climate change policies in an integrated manner, with the potential effects of specific policy measures and technological advances on climate change taken into consideration. Well-known IAMs in climate policy research include the MACRO model (Manne & Richels, 1992), the DICE/RICE model (Nordaus, 1994), the ICAM3 model (Dowlatabadi, 1998), the IMAGE model (Alcamo et al., 1998), the Pizer model (1999), the R&RICE model (Nordhous, 2002), the ETC-RICE model (Buonanno et al., 2003), and the ENTICE model (Popp, 2004). The MACRO model assumes that ESTC takes place in AEEI form. The DICE/RICE model integrates macroeconomics, climate change theory and geography, also assuming ESTC in AEEI form. With the ICAM3 model assuming price- and learning-induced ESCT, Dowlatabadi (1998) empirically analyzes the sensitivity of different economic actors to climate change policies under different technological assumptions. The IMAGE model consists of three subsystems: energy-economy, land-environment, and atmosphere–ocean, assuming that ESTC takes the form of AEEI and is price-induced. Developed by Nordhaus on the basis of the DICE/RICE model, the R&RICE model assumes that ESTC is

R&D-induced. Similar to the R&RICE model, the ETC-RICE model is based on the DICE/RICE model, assuming R&D-induced technological change. The ENTICE model also assumes R&D-induced technological change based on the DICE/RICE model, but unlike the R&RICE and the ETC-RICE model, Popp (2004) assumes that energy R&D investment will only partially crowd out other types of R&D investment.

2.4 This Book's Perspectives

In this chapter, I first summarize R&D-induced endogenous growth theory, including first-generation endogenous growth theory, semi-endogenous growth theory and Schumpeterian growth theory. Then I review, analyze and compare literatures on ESTC in different branches of economics, especially those on the different causes of ESTC under growth theories and climate change economics, including exogenous ESTC, semi-endogenous ESTC and endogenous ESTC. Finally, I enumerate the main methods for empirical studies of ESTC in foreign countries and of energy efficiency and total-factor energy productivity in China, and their current states, including econometric analysis, index decomposition analysis, data envelopment analysis, computable general equilibrium and integrated assessment models. By doing so, I intend to explore the evolution of ESTC theories and their interrelations and illustrate the major approaches to ESTC studies. Through literature review, analysis and comparison, I find ESTC is predicated on two prerequisites. Firstly, the elasticity of substitution between energy and other non-energy factors must be greater than or equal to 1. Secondly, energy-augmenting technological change must be present in the production function. The greater-than-or-equal-to-1 elasticity of substitution crowds out energy input through the effect of factor substitution, while energy-augmenting technological change reduces energy use through increased effective energy input.

One can easily see that ESTC models are rooted in economic growth theory, and that in recent years they have increasingly tended to acknowledge that technological advances are endogenous. However, a large body of literature blurs the line between the effect of increased effective energy input and the effect of factor substitution, and tries to explain energy efficiency gains either by the effect of factor substitution or by the effect of increased effective energy input. It is worthwhile to clarify the basic concept of ESTC, therefore, and distinguish between the effect of increased effective energy input and the effect of factor substitution and their different impacts on energy efficiency. This book begins with defining ESTC, and discusses the existing theories of growth and the basic frameworks for endogenous ESTC in climate change economics as the logical point of departure and theoretical basis (Fig. 2.1), so as to lay a solid foundation for the subsequent chapters.

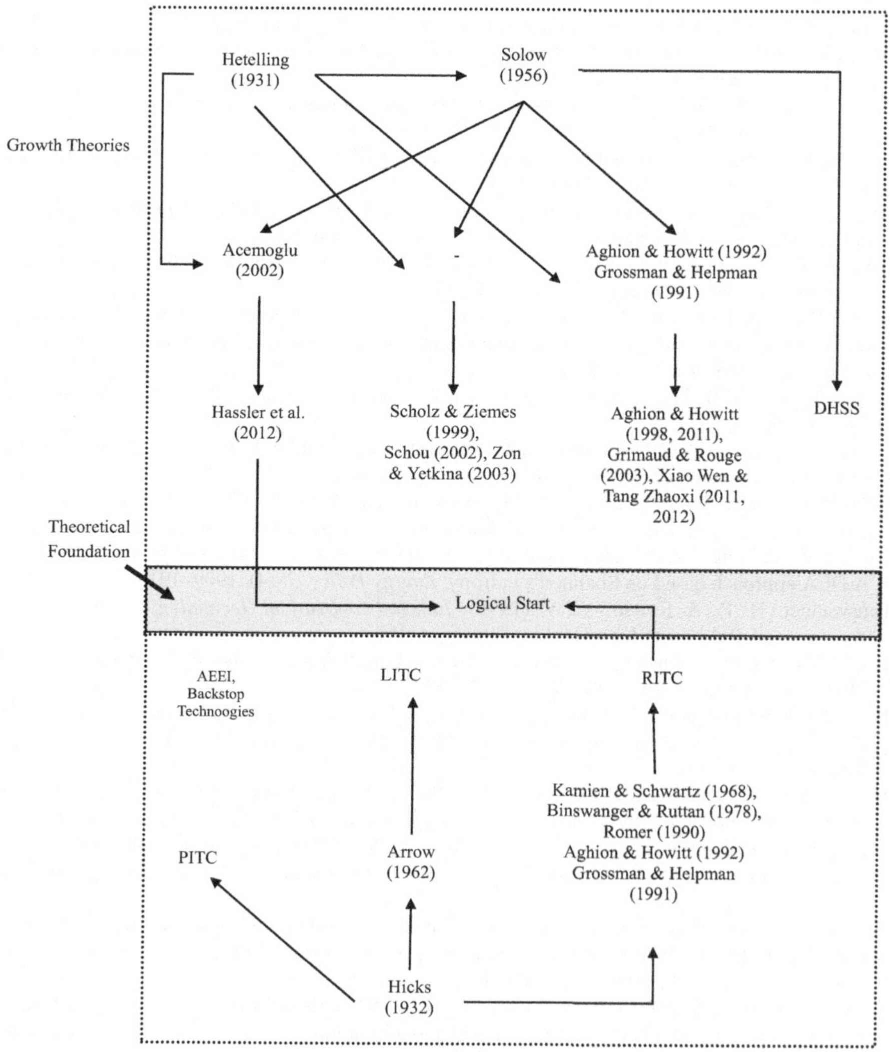

Fig. 2.1 The theoretical foundation and logical point of departure of this book

References

Abadie, L., & Chamorro, J. (2008). European CO_2 prices and carbon capture investment. *Energy Economics, 30*(4), 2992–3015.

Acemoglu, D. (2002). Directed technical change. *Review of Economic Studies, 69*(4), 781–809.

Aghion, P., & Howitt, P. (1992). A model of growth through creative destruction. *Econometrica, 60*(2), 323–351.

Aghion, P., & Howitt, P. (1998). *Endogenous Growth Theory*. MIT Press.

Aghion, P., & Howitt, P. (2011). The economics of growth. Cambridge: MIT Press. Chinese Translation. Beijing: China Renmin University Press.

Ahmad, S. (1966). On the theory of induced invention. *The Economic Journal, 76*(302), 344–357.

Aijun, L. (2010). General equilibrium analysis of China's medium- and long-run energy saving potential. Statistical Research, 03, 45–52.

Alcamo, J., Kreileman, E., & Leemans, R. (1998). *Global change scenarios of the 21st century: Results from the IMAGE 2.1 model*. London: Elsevier Science.

Ang, B. W. (2004). Decomposition analysis for policymaking in energy: Which is the preferred method. *Energy Policy, 32*(9), 1131–1139.

Ang, B. W. (2006). Monitoring changes in economy-wide energy efficiency: From energy–GDP ratio to composite efficiency index. *Energy Policy, 34*(5), 574–582.

Ang, B. W., & Zhang, F. Q. (2000). A survey of index decomposition analysis in energy and environmental studies. *Energy, 25*(12), 1149–1176.

Ang, J. B., & Madson, J. B. (2011). Can second-generation endogenous growth models explain the productivity trends and knowledge production in the Asian Miracle Economies. *The Review of Economics and Statistics, 93*(4), 1360–1373.

Arrow, K. J. (1962). The economic implication of learning by doing. *The Review of Economic Studies, 29*(3), 123–155.

Barker, T., & Köhler, J. (1998). Equity and ecotax reform in the EU: Achieving a 10% reduction in CO_2 emissions using excise duties. *Fiscal Studies, 19*(4), 375–402.

Berry, S., Kortum, S., & Pakes, A. (1996). Environmental change and hedonic cost functions for automobiles. *Proceedings of the National Academy of Sciences, USA, 93*(23), 12731–12738.

Bian, Y. W., & Yang, F. (2010). Resource and environment efficiency analysis of provinces in China: A DEA approach based on Shannon's entropy. *Energy Policy, 38*(4), 1909–1917.

Binswanger, H. P., & Ruttan, V. W. (1978). *Induced innovation: Technology, institution and development*. Baltimore: John Hopkins University Press.

Bo, S. (2007). Energy consumption, structural mutation and Chinese economic growth: 1952–2005. Modern Economic Science, 05, 94–100+127–128.

Bo, S., & Kunrong, S. (2008). Total-factor energy efficiency in segmented Chinese markets: An empirical study based on a super-efficiency DEA model. *The Journal of World Economy, 09*, 49–59.

Böhringer, C., Löschel, A., & Rohde, A. (2002). PACE—Policy analysis based on computable equilibrium. In *Forum für Energiemodelle und Energiewirtschaftliche Systemanalysen in Deutschland, Energiemodelle zum Kernenergieausstieg in Deutschland, Effekte und Wirkungen eines Verzichts auf Strom aus Kernkraftwerken, Umwelt and Ökonomie Bd*. 34, Heidelberg (pp. 436–450).

Bosetti, V., Carraro, C., Duval, R., Sgobbi, A., & Tavoni, M. (2009). The role of R&D and technology diffusion in climate change mitigation: New perspectives using the witch model. Fondazione Eni Enrico Mattei Working Paper No. 200914.

Bosetti, V., Carraro, C., Duval, R., & Tavoni, M. (2011). What should we expect from innovation? A model-based assessment of the environmental and mitigation cost implications of climate-related R&D. *Energy Economics, 33*(6), 1313–1320.

Bosetti, V., & Tavoni, M. (2009). Uncertain R&D, backstop technology and GHGs stabilization. *Energy Economics, 31*(Supplement), S18–S26.

Buonanno, P., Carlo, C., & Marzio, G. (2003). Endogenous induced technical change and the costs of Kyoto. *Journal of Resource and Energy Economics, 25*(1), 11–34.

Burniaux, J., Martin, J. P., Nicoletti, G., & Oliveira-Martins J. (1992). Green: A multi-sector, multi-region general equilibrium model for quantifying the costs of curbing CO_2 emissions: A technical manual. OECD Economics Department Working Paper, No.116.

Capros, P., & Mantzos, L. (2000). Endogenous learning in European Post-Kyoto scenarios: Results from applying the market equilibrium model PRIMES. *International Journal of Global Energy Issue, 14*(1), 249–261.

Carraro, C. (2002). Climate change policy: Models, controversies and strategies. In Tietenberg, T., & Folmer, H. (Eds.) *The International Yearbook of Environmental and Resource Economics*, Oxon: Edward Elgar Publishing Ltd.

Carraro, C., & Galeotti, M. (1997). Economic growth, international competitiveness and environmental protection: R&D and innovation strategies with the WARM model. *Energy Economics, 19*(1), 2–28.

Chang, M., & Yujie, C. (2012). A study of the changes in regional energy efficiency in China from 1995 to 2010—Empirical study of descriptive characteristics and the panel data-based DEA method. *Finance & Trade Economics, 06*, 116–123.

Chauhan, N. S., Mohapatra, P. K. J., & Pandey, K. P. (2006). Improving energy productivity in paddy production through benchmarking: An application of data envelopment analysis. *Energy Conversion and Management, 47*(9–10), 1063–1085.

Chengliang, Y., & Liutang, G. (2009). Schumpeterian growth theory: A literature review. *China Economic Quarterly, 03*, 1163–1196.

Chu, W., Limin, D., & Manhong, S. (2010). Can China achieve energy and emissions reduction goals? A DEA-based assessment and simulation. The Journal of World Economy, 03, 141–160.

Chu, W., & Manhong, S. (2008). Can structural adjustment improve energy efficiency? A study based on Chinese provincial data. *The Journal of World Economy, 11*, 77–85.

Cohen, W. M., & Levinthal, D. A. (1989). Innovation and learning: The two faces of R&D. *The Economic Journal, 99*(397), 569–596.

Cox, A., Trieu, H., Warr, S., & Rolph, C. (1997). Trends in Australian energy intensity. ABARE Research Report No. 1997.5.

Dasgupta, P., & Heal, G. (1974). The optimal depletion of exhaustible resources. *Review of Economic Studies, 41*(5), 3–28.

Dasgupta, P. S., & Heal, G. M. (1980). *Economic theory and exhaustible resources*. Cambridge University Press.

Demin, C., & Rui, Z. (2012). The impact of environmental regulation on total-factor energy efficiency in China—An empirical test based on provincial panel data. *Economic Science, 04*, 49–65.

Dinopoulos, E., & Thompson, P. (1998). Schumpeterian growth without scale effects. *Journal of Economic Growth, 3*(4), 313–335.

Donglan, Z, & Dequn, Z. (2010). A CGE model-based study of the rebound effect in energy efficiency in China. The Journal of Quantitative & Technical Economics, 12, 39–53+66.

Donglin, S., Linhui, W., & Zhiqing, D. (2011). Capital-embodied technological change and its contribution to economic growth (1981–2007). Social Sciences in China, *02*, 91–106+222.

Dowlatabadi, H. (1998). Sensitivity of climate change mitigation estimates to assumption about technical change. *Energy Economics, 20*(5–6), 473–493.

Fangming, Q., & Bo, S. (2010). Index decomposition analysis of changes in energy intensity in China: 1980–2007. *Journal of Beijing Normal University (social Sciences), 02*, 118–126.

Färe, R., Grosskopf, S., Fukuyama, H., & Margaritis, D. (2011). DEA and endogenous technological change. *European Journal of Operational Research, 210*(1), 349–358.

Feng, S., & Lili, W. (2012). Trends in China's energy intensity and inter-provincial differences. *Resources Science, 01*, 13–19.

Gerlagh, R. (2008). A climate-change policy induced shift from innovation in carbon-energy production to carbon-energy savings. *Energy Economics, 30*(2), 425–448.

Gerlagh, R., & Lise, W. (2005). Carbon taxes: A drop in the ocean, or a drop that erodes the stone? The effect of carbon taxes on technological change. *Journal of Ecological Economics, 54*(2–3), 241–260.

Goulder, L. H., & Mathai, K. (2000). Optimal CO_2 abatement in the presence of induced technological change. *Journal of Environmental Economics and Management, 21*(1), 1–38.

Goulder, L. H., & Schneider, S. H. (1999). Induced technological change and the attractiveness of CO_2 abatement policies. *Resource and Energy Economics, 21*(3), 211–253.

Grimaud, A., & Rouge, L. (2003). Non-renewable resources and growth with vertical innovation optimum, equilibrium and economic policies. *Journal of Environmental Economics and Management, 45*, 433–453.

Grossman, G. M., & Helpman, E. (1991). Quality ladders in the theory of growth. *The Review of Economic Studies, 58*(1), 43–61.

Grossman, G. M., & Helpman, E. (1994). Endogenous innovation in the theory of growth. *Journal of Economic Perspectives, 8*(1), 23–44.

Guo Juan, Guo Chonghui, & Ling Yu (2010). Energy rebound effect in Chinese industry. *The Journal of Quantitative & Technical Economics*, 11, 114-126.

Guozhang, L., & Zongjie, H. (2009). Total-Factor Energy Efficiency, Convergence and its Drivers—An empirical analysis based on provincial panel data 1995–2006. *Economic Review, 06*, 101–109.

Ha, J., & Howitt, P. (2007). Accounting for trends in productivity and R&D: A Schumpeterian critique of semi-endogenous growth theory. *Journal of Money, Credit and Banking, 39*(4), 733–774.

Haining, W., & Yuanyuan, C. (2010). Industrial agglomeration effect and industrial energy efficiency—Based on an empirical study of 25 industrial subsectors in China. *Journal of Finance and Economics, 09*, 69–79.

Harris, J., & Thorpe, S. (2000). Trends in Australian energy intensity: 1973–74 to 1997–98. ABARE Research Report No. 2000.13.

Hassler, J., Krusell, P., & Olovsson, C. (2012). Energy-saving technical change. NBER Working Paper No.18456.

Hicks, J. R. (1932). *The Theory of Wages*. Macmillan.

Hotelling, H. (1931). The economics of exhaustible resources. *Journal of Political Economy, 39*(2), 137–175.

Houshyar, E., SheikhDavoodi, M. J., & Nassiri, S. M. (2010). Energy efficiency for wheat production using data envelopment analysis (DEA) Technique. *Journal of Agricultural Technology, 6*(4), 663–672.

Howitt, P. (1999). Steady endogenous growth with population and R&D inputs growing. *Journal of Political Economy, 107*(5), 715–730.

Hu, J. L., & Kao, C. H. (2007). Efficient energy-saving targets for APEC economies. *Energy Policy, 35*(1), 373–382.

Hu, J. L., & Wang, S. C. (2006). Total-factor energy efficiency of regions in China. *Energy Policy, 34*(17), 3206–3217.

Hui Jin & Dale W. Jorgenson (2010), Econometric modeling of technical change. *Journal of Econometrics*, Volume 157(2), 205-219.

IPCC. (2000). Emissions scenarios. Cambridge: Cambridge University Press. Available at http://www.ipcc.ch/ipccreports/sres/emission/index.php?idp=0.

Jacoby, H. D., Reilly, J. M., McFarland, J. R., & Paltsev, S. (2006). Technology and technical change in the MIT EPPA model. *Energy Economics, 28*(5–6), 610–631.

Jakeman, G., Hanslow, K., Hinchy, M., Fisher, B., & Woffenden, K. (2004). Induced innovations and climate change policy. *Energy Economics, 26*(6), 937–960.

Jian, Z. (2008). A study of the dynamic evolution of regional growth and energy efficiency improvement in China—Empirical analysis based on a provincial panel data cointegration model. *The Journal of Quantitative & Technical Economics, 09*, 3–16.

Jones, C. I. (1995a). Time series tests of endogenous growth models. *The Quarterly Journal of Economics, 110*(2), 495–525.

Jones, C. I. (1995b). R&D-based models of economic growth. *Journal of Political Economy, 103*(4), 759–784.

Jones, C. I. (1999). Growth: With or without scale effects. *The American Economic Review, 89*(2), 139–144.

Jorgenson, D. W., & Wilcoxen, P. J. (1993). Reducing US carbon emissions: An econometric general equilibrium assessment. *Resource and Energy Economics, 15*(1), 7–25.

Kamien, M. I., & Schwartz, N. L. (1968). Optimal "induced" technical change. *Econometrica, 36*(1), 1–17.

Keliang, W., Baochen, Y., & Li, Y. (2010). Economic efficiency of energy utilization, environmental performance and energy saving and emissions reduction potentials in China. *Business Management Journal, 10*, 1–9.

Kelley, A. (1988). Economic consequences of population change in the third world. *Journal of Economic Literature, 26*(4), 1685–1728.

Kennedy, C. (1964). Induced bias in innovation and the theory of distribution. *The Economic Journal, 74*(295), 541–547.

Kocherlakota, N., & Yi, K. (1997). Is there endogenous long-run growth? Evidence from United States and United Kingdom. *Journal of Monetary Credit and Banking, 29*(2), 235–262.

Kortum, S. S. (1997). Research, patenting, and technological change. *Econometrica, 56*(6), 1389–1420.

Kouvaritakis, N., Soria, A., & Isoard, S. (2000a). Modelling energy technology dynamics: Methodology for adaptive expectations models with learning by doing and learning by searching. *International Journal of Global Energy Issues, 14*(1), 104–115.

Kouvaritakis, N., Soria, A., Isoard, S., & Thonet, C. (2000b). Endogenous learning in World Post-Kyoto Scenarios: Application of the POLES model under adaptive expectations. *International Journal of Global Energy Issue, 14*(1), 222–248.

Li, C. W. (2000). Endogenous versus semi-endogenous growth in a two R&D-sector model. The Economic Journal, *110*(462), C109–C122.

Ling, T., & Zhenglin, Y. (2009). Energy efficiency and transformation of the industrial economy—An empirical study based on China's Industrial Data 1998–2007. *The Journal of Quantitative & Technical Economics, 10*, 34–48.

Linn, J. (2008). Energy prices and the adoption of energy saving technology. *The Economic Journal, 118*(533), 1986–2012.

Löschel, A., & Schymura, M. (2013). Modeling technological change in economic models of climate change: A survey. ZEW Working Paper No. 13-007.

Löschel, A. (2002). Technological change in economic models of environmental policy: A survey. *Ecological Economics, 43*(2), 105–126.

Löschel, A., & Otto, V. M. (2009). Technological uncertainty and cost effectiveness of CO2 emission reduction. *Energy Economics, 31*(S1), S4–S17.

Loulou, R., Remne, U., Kanudia, A., Lehtila, A., & Goldstein G. (2005). Documentation for the TIMES model: Part 1[R/OL]. http://www.iea-etsap.org/web/Documentation.asp.

MacCracken, C., Edmonds, J., Son, K., & Sands, R. (1999). The economics of the Kyoto protocol. Energy Journal, *20*(Special Issue), 25–71.

Madsen, J. B. (2008). Semi-endogenous versus Schumpeterian growth models: Testing the knowledge production function using international data. *Journal of Economic Growth, 13*(1), 1–26.

Manne, A., & Richels, R. (1992). *Buying greenhouse insurance: The economic costs of CO_2 emission limits*. Cambridge: MIT Press.

McKibbin, W. J., & Wilcoxen, J. P. (1992). G-Cubec: A dynamic multi-sector general equilibrium model of the global economy (Quantifying the Cost of Curbing CO_2 Emissions). Brookings Institution Working Paper No. 98.

Meadows, D. H., Meadows, D. I., Randers, J. et al. (1997). The limits to growth: A report to the club of Rome. New South Wales: Universe Books, 1974, Chinese Translation. Changchun: Jilin People's Publishing House.

Messner, S., & Schrattenholzer, L. (2000). MESSAGE-MACRO: Link an energy supply model with a macroeconomic module and solving it iteratively. *Energy Journal, 25*, 267–282.

Mukherjee, K. (2008). Energy use efficiency in U.S. manufacturing: A nonparametric analysis. Energy Economics, *30*(1), 76–96.

Myers, J., & Nakamura, L. (1978). *Saving energy in manufacturing*. Ballinger.

Newell, R. G., Jaffe, A. B., & Stavins, R. N. (1999). The induced innovation hypothesis and energy-saving technological change. *The Quarterly Journal of Economics, 114*(3), 941–975.

Nordhaus, W. (2002). Modeling induced innovation in climate change policy. In Grubler, Nakicenovic, & Nordhause (Eds.) *Technological change and the environment*. Washington DC: Resources for Future Press.

Nordhaus, W. (1992). An optimal transition path for controlling greenhouse gases. *Science, 258*(5086), 1315–1319.

Nordhaus, W. (1994). *Managingthe global commons: The economics of climate change*. MIT Press.

Onut, S., & Soner, S. (2006). Energy efficiency assessment for the Antalya Region Hotels in Turkey. *Energy and Buildings, 38*(8), 964–971.

Oravets, M., & Dowlatabadi, H. (1996). Modeling US long-term energy intensity: An exploration of technological change. Department of Engineering and Public Policy Working Paper, Caenegie Mellon University.

Otto, V. M., Löschel, A., & Reilly, J. (2008). Directed technical change and differentiation of climate policy. *Energy Economics, 30*(6), 2855–2878.

Pakes, A., Berry, S., & Levinsohn, J. A. (1993). Applications and limitations of some recent advances in empirical industrial organization: Prices indexes and the analysis of environmental change. *American Economic Review, 83*(2), 240–246.

Pengfei, Y., & Jilin, W. (2011). General characteristics of energy consumption and economic fluctuation in energy price hike scenarios. *Statistical Research, 09*, 57–65.

Peretto, P. F. (1998). Technological change and population growth. *Journal of Economic Growth, 3*(4), 283–311.

Peretto, P., & Smulders, S. (2002). Technological distance, growth and scale effects. *The Economic Journal, 112*(481), 603–624.

Petchey, R. (2010). End use energy intensity in the Australian economy. ABARE Research Report No. 2010.02.

Pizer, W. (1999). The optimal choice of climate change policy in the presences of uncertainty. *Resource and Energy Economics, 21*(3–4), 255–287.

Popp, D. (2002). Induced innovation and energy prices. *The American Economic Review, 92*(1), 160–180.

Popp, D. (2004). ENTICE: Endogenous technological change in the DICE model of global warming. *Journal of Environmental Economics and Management, 48*(1), 742–768.

Popp, D. (2005). ENTICE-BR: The effects of backstop technology R&D on climate policy models. *Energy Economics, 28*(2), 188–222.

Popp, D., Newell, R. G., & Jaffe, A. B.(2009). Energy, the environment, and technological change. NBER Working Paper, No. 14832.

Qiaosheng, W. (2010). Changes in energy intensity and their drivers in Chinese industrialization—A Fisher IDA-based empirical analysis. Economic Theory and Business Management, 05, 44–50.

Qiubin, W. (2010). Industrial energy efficiency and industrial structure optimization and upgrading—An empirical study based on provincial panel data 2000–2006. *The Journal of Quantitative & Technical Economics, 10*, 49–63.

Richels, R. G., & Blanford, G. J. (2008). The value of technological advance in decarbonizing the U.S. economy. Energy Economics, *30*(6), 2930–2946.

Romer, P. M. (1990). Endogenous technological change. *Journal of Political Economy, 98*(5), S71–S102.

Romer, P. M. (1993). Idea gaps and object gaps in economic development. *Journal of Monetary Economics, 32*(3), 543–573.

Rosenberg, N. (1982). *Inside the black box: Technology and economics*. Cambridge University Press.

Ruizhi, P., Luxian, W., & Quan, Z. (2009). The factor effect, structural effect and technical effect in total-factor energy efficiency in Chinese industry in transition. Modern Economic Science, 05, 21–27+124–125.

Sandu, S., & Syed, A. (2008). Trends in energy intensity in Australian industry. ABARE Research Report No.2008.08.

Sandu, S., & Petchey, R. (2009). End use energy intensity in the Australian economy. ABARE Research Report No.2009.03.

Scholz, C. M., & Ziemes, G. (1999). Exhaustible resources, monopolistic competition and endogenous growth. *Environmental and Resource Economics, 13*, 169–185.

Schou, P. (2002). When environmental policy is superfluous: Growth and polluting resources. *Scandinavian Journal of Economics, 104*, 605–620.

Segerstrom, P. S. (1998). Endogenous growth without scale effects. *The American Economic Review, 88*(5), 1290–1310.

Segerstrom, P. S., Anant, T. C. A., & Dinopoulos, E. (1990). A Schumpeterian model of the product life cycle. *The American Economic Review, 80*(5), 1077–1091.

Shaojun, Z, & Dongfang, L. (2009). Disintegration of production and energy efficiency—Empirical studies based on panel data on Chinese industry. China Industrial Economics, 02, 66–75.

Sheng, Z., & Dengshi, H. (2009). Energy consumption, economic growth and energy efficiency in China—Based on Empirical Studies 1980–2007[J]. *The Journal of Quantitative & Technical Economics, 08*, 17–28.

Shixiang, L., & Jinhua, C. (2008). Assessment of energy efficiency in China and analysis of its drivers. *Statistical Research, 10*, 18–27.

Shiyuan, P., & Jinchuan, S. (2002). The theory of endogenous growth: A literature review. *China Economic Quarterly, 03*, 753–786.

Shuai, S., Lili, Y., & Jianhua, C. (2010). A study of the drivers of carbon emissions from industrial energy consumption—STIRPAT model-based empirical analysis of dynamic panel data on Shanghai industrial subsectors. *Journal of Finance and Economics, 11*, 16–27.

Smulders, S., & Nooij, M. (2003). The impact of energy conservation on technology and economic growth. *Resource and Energy Economics, 25*(1), 59–79.

Solow, R. M. (1974). Intergenerational equity and exhaustible resources. *Review of Economic Studies, 41*(5), 29–45.

Stiglitz, J. E. (1974a). Growth with exhaustible natural resources: Efficient and optimal growth paths. *Review of Economic Studies, 41*(5), 123–137.

Stiglitz, J. E. (1974b). Growth with exhaustible natural resources: The competitive economy. *Review of Economic Studies, 41*(5), 139–152.

Sue Wing, I. (2001). Induced technical change in computable general equilibrium models for climate-change policy analysis. PhD dissertation, Massachusetts Institute of Technology, Engineering Systems Division.

Sue Wing, I. (2003). Induced technical change and the cost of climate policy. MIT Joint Program Technical Note No. 6.

Sue, W. I. (2006). Representing induced technological change in models for climate policy. *Energy Economics, 28*(5–6), 539–562.

Sue, W. I. (2008). Explaining the declining energy intensity of the US economy. *Resource and Energy Economics, 30*(1), 21–49.

Tedesco, L., & Thorpe, S. (2003). Trends in Australian energy intensity: 1973–74 to 2000–01. ABARE Research Report No. 2003.9.

Temple, J. (2003). The long-run implication of growth theories. *Journal of Economic Surveys, 17*(3), 497–510.

Thompson, P. (2012). The relationship between unit cost and cumulative quantity and the evidence for organizational learning-by-doing. *The Journal of Economic Perspectives, 26*(3), 203–224.

van der Werf E., (2008). Production functions for climate policy modeling: An empirical analysis. *Energy Economics*, 30(6), 2964-2979.

van Zon, A., & Yetkiner, I. H. (2003). An endogenous growth model with embodied energy-saving technical change. *Resource and Energy Economics, 25*, 81–103.

Wang, K., Wang, C., & Chen, J. (2009). Analysis of the economic impact of different Chinese climate policy options based on a CGE model incorporating endogenous technological change. *Energy Policy, 37*(8), 2930–2940.

Wei, W., & Xubo, H. (2012). R&D subsidy policy options under the dual goals of energy and emissions reduction and economic growth—A DCGE-based scenario analysis. Modern Economic Science, 06, 70–80+124–125.

Wei, W., & Xubo, H. (2013). Energy and emissions reduction, R&D subsidies and sustainable growth—A DCGE-based scenario analysis. *Business Management Journal, 11*, 1–12.

Wen, X., & Zhaoxi, T. (2011). Energy restraints, technological change and sustainable development—A framework of analysis based on the improvement of intermediate goods quality. *Economic Theory and Business Management, 01*, 87–94.

Wen, X., & Zhaoxi, T. (2012). Renewable energy, quality of intermediate goods and sustainable development. *The World Economy, 02*, 143–159.

Wilson, B., Trieu, L. H., & Bowen, B. (1994). Energy efficiency trends in Australia. *Energy Policy, 22*(4), 287–295.

Xiao'e, Q. (2009). Decomposition of changes in inter-provincial total-factor energy efficiency in China: An empirical study based on Malmquist index. The Journal of Quantitative & Technical Economics, 08, 29–43.

Xiao, W., & Xiao'e, Q. (2010). Total-factor energy efficiency in Chinese manufacturing—Based on an empirical study of 28 manufacturing subsectors. Modern Economic Science, 02, 20–25+124–125.

Xiaoling, Y., Baoshan, Z., & Wanping, Y. (2009). A study of total-factor energy efficiency in China based on environmental pollution. *China Industrial Economics, 02*, 76–86.

Xiaoming, G., & Zongyi, Z. (2012). Drivers of energy intensity in the Chinese transport industry. *Journal of Industrial Engineering and Engineering Management, 04*, 90–99.

Xin, Y., & Xiying, L. (2010). The optimal carbon tax of China from a growth perspective. *Economic Research Journal, 11*, 48–58.

Xin, Y., Zhujun, J., & Jianghua, L. (2011). Reforming fossil fuel subsidies can support development of clean energy. *Journal of Financial Research, 03*, 184–197.

Yijun, Y., Lili, G., & Jia, S. (2012). Structure, technology, management and energy efficiency—Analysis based on China's provincial panel data 2000–2010. *China Industrial Economics, 07*, 18–30.

Young, A. (1998). Growth without scale effect. *Journal of Political Economy, 106*(1), 41–63.

Yuanlong, L., & Wencong, L. (2011). Rebounds of macro-energy consumption due to efforts to improve energy efficiency in productive sectors. China Population, Resources and Environment, 11, 44–49.

Yuming, W., & Lin, J. (2009). Stochastic convergence of regional energy efficiency in China. Economic Science, 06, 40–49.

Zhiyun, Z., Bingyang, L., Qingwang, G., & Junxue, J. (2007). Dynamic integration of capital accumulation and technological change: A typical fact of Chinese economic growth. Economic Research Journal, 11, 18–31.

Zhongdong, Y. (2010). Drivers of energy efficiency in Chinese manufacturing: Business cycles and industrialization in the heavy chemical industry. *Statistical Research, 10*, 33–39.

Zhou, P., Ang, B. W., & Poh, K. L. (2008). A survey of data envelopment analysis in energy and environmental studies. *European Journal of Operational Research, 189*(1), 1–18.

Zha Jianping, Tang Fangfang & Fu Hao (2011). Energy consumption, Carbon emissions and industrial growth—An empirical analysis from a decoupling perspective. *Contemporary Economic Science*, 06, 81-89+125.

Chapter 3
The Analytical Framework for Understanding R&D Heterogeneity-Based ESTC in Industry

Since an economy is a grand system consisting of numerous economic actors, social institutions and the natural environment, and complex interrelationships between them, economists are compelled to form various economic concepts in one way or another as subsystems constituting a structure, which is often called a framework of economic analysis (Yang Xiaokai & Huang Youguang, 1999). Under such a framework, as long as a reality conforms to the hypothetical economic conditions and conduct patterns proposed by an economic theory, one can draw logical conclusions from the theory and provide solutions to various problems in question (Tian Guoqiang, 2005). According to Yang Xiaokai and Huang Youguang (1999) and Yang Xiaokai (2003), in the course of theorizing, the process of conceptualization is sometimes more important than the resultant concept itself. As noted in Chap. 2 Literature Review, not only are there multiple analytical frameworks for endogenous ESTC in the existing literature, but they define ESTC in much different ways. My first task in this chapter, therefore, is to clarify the basic concept of ESTC. On that basis, I differentiate between the effect of energy R&D investment and that of non-energy R&D investment on energy efficiency, and analyze R&D heterogeneity and energy efficiency gains under one framework.

3.1 Basic Concept and Characteristics

3.1.1 Energy-Saving Technological Change

(1) Technological Change

In Joseph A. Schumpeter's view Schumpeter (1990), technological change occurs in three phases: invention, innovation and diffusion. The Intergovernmental Panel on Climate Change (IPCC) (2000) describes technological change as including the following five stages of development: (1) scientific discovery helping add to the

X. He, *The Endogenous Energy-Saving Technological Change in China's Industrial Sector*, https://doi.org/10.1007/978-981-19-7485-4_3

stock of knowledge; (2) invention, the attempt to apply existing knowledge to practical ends; (3) innovation, the implementation of an invention for the first time; (4) improvement of technical features and reduction of costs of using a technology; and (5) diffusion of innovation, often accompanied by technological advancement. Theories of economic growth generally define technological change as shifts in the production function, which incorporate the definitions of technological change given by Schumpeter and IPCC (2000) respectively. In developing a framework of analysis, theorists tend to divide technological change into two categories: exogenous or endogenous technological change depending on whether it is driven by the optimal behavior of economic actors; and embodied or non-embodied technological change depending on whether it only results from additional factors of production. Because the existing theory of embodied technological change focuses on the improvement of capital quality, it can be called capital-embodied technological change. While capital-embodied technological change only affects the quality of new capital, continuous investment is necessary to improve capital quality so as to achieve technological change. On the other hand, non-embodied technological change is attained through increasing effective input factors.

(2) Energy Productivity and Energy Efficiency

Productivity means the ratio of output to the total input required to produce it (b; Wei Chu & Shen Manhong, 2007a). Depending on the number of factors involved, it can be either single-factor productivity or multi-factor or total-factor productivity. Single-factor productivity is a measure of output against a specific input, while multi-factor or total-factor productivity is the ratio between output and the weighted average of inputs (the weight is the elasticity of output of each input). According to Farrell (1957), efficiency includes technical efficiency and allocative efficiency. Technical efficiency means the capacity to optimally use available resources, namely, to maximize output with other inputs held fixed, or to minimize inputs at a given level of output (Lovell, 1995). Allocative efficiency refers to the capacity to arrive at the optimal combination of input factors and output at certain factor prices. Wei Chu and Shen Manhong (2007a, b) and Yang Qingwen (2012) summarize the differences between productivity and efficiency as in Table 3.1.

(1) Energy Productivity

It follows from the definition of productivity that energy productivity is the ratio between output and energy input required to achieve it (Wei Chu, 2007; Wei Chu & Shen Manhong, 2007a, b). If output is measured by gross domestic product (GDP), then energy productivity is:

$$\text{Energy Productivity} = \frac{\text{GDP}}{\text{Eenrgy Input}} \tag{3.1}$$

Equation (3.1) calculates single-factor energy productivity as it includes only one factor of production (i.e., energy). This method of calculation, which is found in most of the literature, renders energy productivity and energy intensity reciprocal. But it may give short shrift to other non-energy factors that may contribute to greater energy efficiency, such as the quality of capital. If the production process involves factors of production in addition to energy, the result of weighted calculation is multi-factor or total-factor energy productivity.

(2) Energy Efficiency

Energy efficiency is a generic term, and there are multiple measures of energy efficiency (Patterson, 1996). In general, energy efficiency improves when less energy input produces the same quantity of services or output, or the same energy input yields more services or output (Wei & Shen, 2007; Wei Chu & Shen Manhong, 2007a, b; Yang Qingwen, 2012). According to Patterson (1996), energy efficiency can be defined as:

$$\text{Energy Efficiency} = \frac{\text{Output or Services}}{\text{Energy Input}} \tag{3.2}$$

Based on different measurements and units for output or services and the quantity of energy input, Patterson (1996) subdivides energy efficiency into four types of indicator: thermodynamic, physical-thermodynamic, economic-thermodynamic, and economic. Following Patterson (1996), Tanaka (2008) likewise defines four types of energy efficiency indicator respectively measuring energy output/energy input, energy intensity, energy consumption, and energy technology diffusion. These indicators of energy efficiency, proposed either by Patterson (1996) or by Tanaka (2008), strictly speaking, are still measures of energy productivity (Please compare (3.1) and (3.2)).

(3) Energy-Saving Technological Change

(1) Basic Concept

As observed in Chap. 2 Literature Review, no agreement has been reached in the existing literature on the basic concept of ESTC. Most of the literature approaches ESTC in either of two ways. One is to define ESTC as changes in energy productivity measured by various methods, regardless of what has brought about those changes. For example, the relevant literature in climate change economics defines energy-related technology change in terms of changes in energy intensity: a positive change in energy intensity means energy-utilizing technology change, and a negative change in energy intensity indicates ESTC (Sue Wing, 2001, 2003; Wang et al., 2009). The other is to introduce energy into growth models as a factor of production. If, because of a certain type of technological change in the model (whether it is exogenous or endogenous, and whether it is neutral or non-neutral), the scarcity of energy does not stunt economic growth, one can say that there has occurred ESTC. A comparison of the above two ways of defining ESTC shows that the first, as adopted by climate change economists, is an ad-hoc method, which can be incorporated in the second method. The biggest defect of the ad-hoc method is that it defines ESTC only through analysis of numerical simulation, without regard to any real reasons behind it. As the existing literature reveals, the chief problem with the second method is a lack of clarity, leading to most of the literature focusing on ESTC ascribed to better quality of capital, and ignoring the potential impact of increased effective energy input on ESTC. Overall, the first definition of ESTC overlooks the effect of other non-energy factors on energy productivity, while the second is overly concerned with the effect of substitution of other factors (especially capital quality) for energy, without considering energy savings from additional effective energy input (This is in stark contrast to the problem with energy productivity, which focuses on energy alone but shuns other factors). What the two definitions have in common is that they both lose sight of the possible role of increased effective energy input in the improvement of energy efficiency. In this sense, I find no satisfactory definition of ESTC in the existing literature.

Hassler et al. (2012) attempt to aid the two defective definitions by assuming R&D heterogeneity in an economy (which means that one R&D sector produces capital/labor-augmenting technology change, whereas the other makes energy-augmenting technological change, with both sectors having the same input factors), and criticizing the theoretical logic of endogenous ESTC theory that neglects increases in effective energy input. Hassler et al.'s (2012) revisions are reflected in the following two assumptions: (1) there exists energy-augmenting technological change in an economy; (2) capital-labor and energy are fully complementary to each other. The revisions enable energy-augmenting technological change in Hassler et al. (2012) to depict improvements in energy efficiency through greater effective energy input. However, their assumption that capital-labor and energy are fully complementary to each other is so strong that it leads to a linear relationship between energy intensity and ESTC. Moreover, giving no thought to the effect of other non-energy factors (such as capital quality) on energy efficiency is likely to degrade ESTC into single-factor energy productivity. It is precisely because Hassler et al. (2012) fail to combine the effects of factor substitution and increased effective energy input that ESTC as defined by Hassler et al. (2012) is incapable of comprehensively measuring energy efficiency. Since both the elasticity of substitution between energy and other factors and energy-augmenting technological change contribute to ESTC, in order to do justice to the effects of factor substitution and increased effective energy input on energy efficiency, I propose to define ESTC as follows.

Definition 4.1 Let $\widetilde{F}(\cdot)$ be the total production function incorporating capital, labor and energy, and $F(\cdot)$ be the production function including the foregoing three factors with constant returns to scale (CRS). $Q(t)$ is an aggregation of capital, labor and energy-augmenting technological change, i.e., $Q(t) = \{Q_K(t), Q_L(t), Q_E(t)\}$; $Q^{KL}(t)$ is an aggregation of capital and labor-augmenting technological change, i.e., $Q^{KL}(t) = \{Q_K(t), Q_L(t)\}$; and σ is an aggregation of the elasticities of substitution between various factors, i.e., $\sigma = \{\sigma_{K,L}, \sigma_{K,E}, \sigma_{L,E}, \sigma_{KL,E}, \sigma_{KE,L}, \sigma_{LE,K}, \sigma_{KLE}\}$. If (3.3) is valid, then Q_E in (3.3) is ESTC:

$$Y(t) = \widetilde{F}(K(t), L(t), E(t), Q(t), \sigma) = F(K(t), L(t), Q_E(t)E(t), Q^{KL}(t), \sigma) \quad (3.3)$$

Table 3.1 Comparison of productivity and efficiency

	Productivity	Efficiency
Definition	Ratio of output to inputs required	Capacity to optimally use resources
Equation	Single-factor productivity: output/specific input Total-factor productivity: output/weighted average of inputs	Input perspective: optimal input/actual input Output perspective: actual output/optimal output
Value	$[0, +\infty)$	[0,1]
Comparison Criteria	Horizontal comparison of different economic actors, or vertical comparison within an economic agent	1 means efficiency, 0 denotes inefficiency
Is there any dimension?	Yes, it depends on units of input and output	None
Strengths	It is easy to calculate, and is fit for comparison of economic actors with large gaps in technical efficiency	It measures to what extent inputs are utilized in actual production, and inclines to embody efficiency factors
Weaknesses	It neglects the effect of other inputs on output	It is complicated to calculate, and different calculation methods yield different values of efficiency. And it is hard to measure the utilization efficiency of individual inputs

Sources Wei Chu and Shen Manhong (2007a, b), Yang Qingwen (2012)

As $F(\cdot)$ has constant returns to scale, the equation would be simpler if $Q^{KL}(t)$ and σ are dropped. Dividing both sides of (3.3) by $L(t)$ would yield:

$$\overline{Y}(t) = F\left(\overline{K}(t), 1, Q_E(t)\overline{E}(t)\right) = \overline{F}\left(\overline{k}(t), Q_E(t)\overline{E}(t)\right) \tag{3.4}$$

where $\bar{\cdot}$ is a per-worker variable. Under the concept of Harrod-neutral technological change, given a certain output-capital ratio $\frac{\overline{Y}(t)}{\overline{K}(t)}$, $\frac{\partial\overline{Y}(t)/\partial\overline{K}(t)}{\partial\overline{Y}(t)/\partial\overline{E}(t)}$, will remain unchanged before and after ESTC. On the one hand, as ESTC increases the marginal output of energy (per-worker), ESTC will exert an influence on economic growth as if "effective" energy input were added. For this reason, ESTC is also called energy-augmenting technological change. In fact, as shown in (3.4), ESTC does not sway the marginal output of capital per-worker, and the key to $\frac{\partial\overline{Y}(t)/\partial\overline{K}(t)}{\partial\overline{Y}(t)/\partial\overline{E}(t)}$ remaining constant before and after ESTC lies in the fact that the output-energy ratio $\frac{\overline{Y}(t)}{\overline{E}(t)}$ increases with the capital-energy ratio $\frac{\overline{K}(t)}{\overline{E}(t)}$ proportionally. On the other hand, ESTC also affects the use of other non-energy factors by changing the marginal productivity of energy (per-worker).

Table 3.2 Measures of ESTC in most of the literature

Literature	Measure	Equation
Binswanger & Ruttan (1978) Jorgenson & Wilcoxen (1993) Popp (2002)	Changes in energy intensity	$\frac{E(t+1)/Y(t+1)}{E(t)/Y(t)}$
Newell et al. (1999)	Polynomial AEEI	$a_0 + a_1 t + a_2 t^2 +$ $a_3 \ln q_t + a_4 \ln p_t + a_5 s$
Sue Wing (2001)	Arithmetic mean of energy intensity changes	$\frac{1}{T-1} \times \sum_{t=1}^{T-1} \left[\frac{E(t+1)/Y(t+1)}{E(t)/Y(t)} - 1 \right]$
Hassler et al. (2012)	Energy-augmenting technological change	$Q_E(t)$ in (3.3) under certain parameters

Note In Polynomial AEEI, t is time, q is the total load of products, p is the relative price of energy, and s is the energy efficiency standard

Table 3.2 summarizes the different measures of ESTC in the existing literature, including changes in energy intensity, polynomial AEEI, the arithmetic mean of energy intensity changes, and energy-augmenting technological change. Measuring ESTC by changes in energy intensity, polynomial AEEI and the arithmetic mean of energy intensity changes may give rise to two problems. Firstly, one would have difficulty in systematically linking the concept of ESTC with the existing theories of economic growth to build a framework of analysis and a formal model incorporating multiple economic actors. Secondly, the three measures fail to take the substitution of other non-energy factors for energy into consideration as much as they should.

(2) Classification

Table 3.3 shows the classification of ESTC. As (3.3) indicates, when $Q(t) = \{Q_K(t), 1, Q_E(t)\}$, and $\sigma_{K,L} = 1$ and $\sigma_{KL,E} = 0$, (3.3) collapses into the production function (2.15) in the final goods producing sectors in Hassler et al. (2012). When $Q(t) = \{Q_K(t), 1, Q_E(t)\}$, and $\sigma_{KLE} = 1$, (3.3) becomes the production function (2.14) in the final goods producing sectors in the framework for embodied technological change.

Table 3.3 Classification of ESTC

	Embodied technological change	Non-embodied technological change
Exogenous		DHSS
Endogenous	Scholz and Ziemes (1999), Schou (2002), Zon and Yetkina (2003), Aghion and Howitt (1998, 2011), Grimaud and Rouge (2003), Xiao Wen and Tang Zhaoxi (2011, 2012)	Hassler et al. (2012)

(3) Comparison

ESTC in Definition 4.1 and energy productivity have the same advantages: they are easy to calculate and have definite meanings; both may be used as proxies for improvements in energy efficiency to a certain extent. Still, the concept of ESTC has other advantages over energy productivity. Firstly, it makes allowance for both increased effective energy input and substitution of other factors for energy. Secondly, Definition 4.1, in which endogenous growth theory can be embedded, can explain ESTC's endogenous nature and how it works. The advantages of ESTC can be explained by (3.5):

$$\overline{Y} = \left[\gamma\left(Q_K\overline{K}\right)^{\frac{\sigma_{K,L}-1}{\sigma_{K,L}}} + (1-\gamma)\left(Q_E\overline{E}\right)^{\frac{\sigma_{K,L}-1}{\sigma_{K,L}}}\right]^{\frac{\sigma_{K,L}}{\sigma_{K,L}-1}} \tag{3.5}$$

where $\overline{Y}$, $\overline{K}$ and $\overline{E}$ respectively represent output per worker, capital per worker and energy per worker; γ is the share of input factors; $\sigma_{K,L}$ is the elasticity of substitution between capital per worker and energy per worker; and Q_K and Q_E respectively denote capital- and energy-augmenting technological change. Equation (3.5) shows how capital per worker and energy input per worker combine to produce final goods through CES technology. Q_E in (3.5) satisfied the requirements of Definition 4.1. If θ_{EY} is energy as a share of output, then (3.6) holds up:

$$\theta_{EY} \equiv \frac{\partial y}{\partial \overline{E}}\frac{\overline{E}}{\overline{Y}} = \gamma\left(\frac{Q_E\overline{E}}{\overline{Y}}\right)^{\frac{\sigma_{K,L}-1}{\sigma_{K,L}}} \tag{3.6}$$

Through transposition, (3.6) becomes:

$$Q_E = \frac{\overline{Y}}{\overline{E}}\left(\frac{\theta_{EY}}{\gamma}\right)^{\frac{\sigma_{K,L}}{\sigma_{K,L}-1}} \tag{3.7}$$

As shown in (3.7), ESTC is positively correlated with energy productivity and substitution elasticity $\left(\frac{\theta_{EY}}{\gamma} > 0\right)$, and it includes both the effects of factor substitution and increased effective energy input. A simple transposition would change (3.7) to:

$$\frac{\overline{E}}{\overline{Y}} = \frac{1}{Q_E}\left(\frac{\theta_{EY}}{\gamma}\right)^{\frac{\sigma_{K,L}}{\sigma_{K,L}-1}} \tag{3.8}$$

On the right-hand side of (3.8) are two causes of declines in energy intensity: ESTC and changes in energy as a share of output. Since the marginal productivity of the factors and the share of energy are also affected by ESTC, ESTC is at least a requisite for improvements in energy intensity.

Equation (3.5) contains all possible forms of production function. If the elasticity of substitution between capital and energy $\sigma_{K,L}$ is 1, (3.5) reduces to a Cobb–Douglas production function:

$$\overline{Y} = \left(Q_K\overline{K}\right)^{\gamma}\left(Q_E\overline{E}\right)^{1-\gamma} \tag{3.9}$$

As under the Cobb–Douglas production function, capital and energy as percentages of output are constants γ and $1-\gamma$ respectively, neither (3.7) nor (3.8) is applicable when $\sigma_{K,L}$ is 1. Equation (3.9) may be rewritten as:

$$Q_E = \left[\left(\frac{1-\gamma}{\gamma}\right)^{\gamma}\frac{\overline{Y}}{\overline{E}}\left(\frac{MPK}{MPE}\right)^{\gamma}\left(\frac{1}{Q_K}\right)^{\gamma}\right]^{\frac{1}{1-\gamma}} \tag{3.10}$$

where *MPK* and *MPE* are the marginal productivity of capital and energy respectively. Since $0 < \gamma < 1$, (3.10) may also be expressed as:

$$Q_E \sim \frac{\overline{Y}}{\overline{E}}\left(\frac{MPK}{MPE}\right)\left(\frac{1}{Q_K}\right) = \gamma\frac{MPK}{Q_K} \tag{3.11}$$

According to (3.11), ESTC is an increasing function of energy productivity and the marginal productivity of capital, and a decreasing function of the marginal productivity of energy and capital-augmenting technological change. When (3.10) is on an isoquant curve, any increase in *MPK* means a reduction in capital investment, so that with other conditions fixed, ESTC needs to be accelerated to ensure the same level of output. Any increase in Q_K on the isoquant curve means additional effective capital input, maintaining the output level by crowding out energy input.

Since neither *MPK* nor *MPE* can be immediately measured, I simulate, based on (3.7) and (3.8), the relationship between ESTC and energy intensity under different elasticities of substitution, using relevant data on China's industrial sectors. As shown in Fig. 3.1, energy intensity and ESTC change basically in reverse directions. Across all values of substitution elasticity, the substitution effect on ESTC is smaller in energy sectors and high energy-consuming sectors than in industry as a whole and in low energy-consuming sectors. Even with high elasticities of substitution, ESTC is slow in energy sectors and high energy-consuming sectors. The simulation shows that ESTC is fairly susceptible to the substitution effect in low energy-consuming sectors.[1]

[1] Figure 3.1 does not represent non-energy sectors because, according to the interpretation of indicators in the *China Statistical Yearbook*, China's energy sectors only include coal mining and washing, oil and natural gas extraction, oil and coke processing, and the production and distribution of power, heat and gas; energy sectors, except gas production and distribution, are all high energy-consuming sectors. That is to say, non-energy sectors include all high energy-consuming sectors except energy sectors, and all low energy-consuming sectors except the production and distribution of gas. Therefore, the simulated dynamics of energy intensity and ESTC in non-energy sectors under different elasticities of substitution should be between Fig. 3.2a, d.

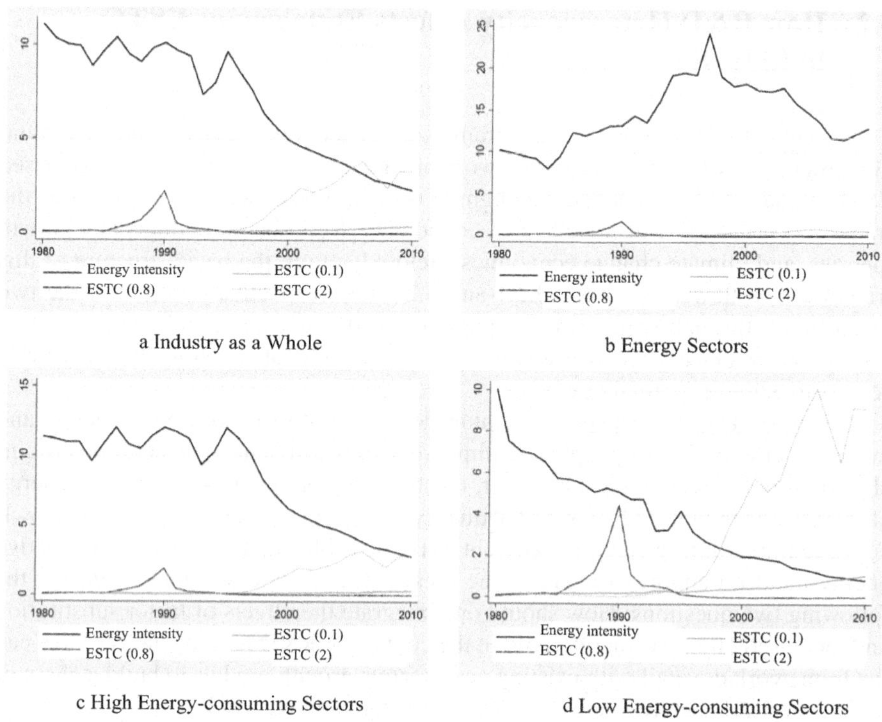

Fig. 3.1 Energy intensity and ESTC in China's industrial sectors under different elasticities of substitution. *Source* The graphs in Fig. 3.1 are calculated based on relevant data on China's industrial sectors (energy intensity, energy share, etc.), as shown in Chap. 5. High energy-consuming and low energy-consuming sectors are distinguished by the energy consumption levels of different industrial sectors in 2010, as detailed in Chap. 5. The figures in brackets denote values of the elasticity of substitution

3.1.2 R&D Heterogeneity

Under neoclassical economic theory, associated with R&D heterogeneity is the fact that different types of R&D investment will affect the marginal productivity of different factors of production, and in turn determine the marginal costs of different factors of production and the choice of factors by manufacturers. As this book is focused on ESTC, I dichotomize R&D heterogeneity in the following chapters in order to remain general. R&D heterogeneity in this book refers to two types of R&D investment: energy R&D investment and non-energy R&D investment. The former is made for the purposes of ESTC, whereas the latter to improve the quality of other non-energy production factors. These two types of R&D investment will go into different R&D sectors, and will influence energy efficiency and the production of final goods.

3.2 How R&D Heterogeneity Affects Endogenous ESTC in Industry

Consistent with the components of a framework of economic analysis and its evaluation criteria, I try to construct a framework of analysis for R&D heterogeneity-based ESTC in industry based on the paradigm of neoclassical economics, drawing on the frameworks and tools of neoclassical economics, R&D-based endogenous growth theories, and climate change economics. Below I set out the basic structure of this framework, followed by Sects. 3.2.1 and 3.2.2 that respectively introduce the two mechanisms through which R&D heterogeneity affects ESTC, and Sect. 3.2.3 that discusses how heterogeneous R&D investments are allocated among different R&D sectors to achieve optimal outcomes.

As demonstrated in Chap. 2 Literature Review, both substitution of factors and increased effective energy input can improve energy efficiency and achieve sustainable growth. Current studies, however, mostly focus on the improvement of energy efficiency attributable to better capital quality or the replacement of energy by knowledge accumulation, and barely shed light on the possible impact of increased effective energy input on energy efficiency. The present framework is bound to answer the following two questions: How should one integrate the effects of factor substitution and increased effective energy input in the context of R&D heterogeneity? How can the framework dovetail with existing endogenous growth models to build a chain of logic from R&D heterogeneity to ESTC to sustainable growth?

To answer the two questions, I adopt the paradigm of neoclassical economics to build an analytical framework, selecting representative households and manufacturers as my objects of study. The representative manufacturers come from four categories: final goods production, intermediate goods production, energy production, and technology R&D. Figure 3.2 shows the structure of my framework for endogenous R&D heterogeneity-based ESTC in industry. The framework is based on four assumptions. First, there is only one final goods producing sector in the economy, which hires labor and buys intermediate goods, energy, and ESTC to organize production for maximization of profits. The production function fits the assumption of the neoclassical production function, and there is a certain degree of substitutability among factors of production. Second, the intermediate goods producing sectors invest capital and purchase R&D products to produce intermediate goods of various grades of quality, and they share a Dixit-Stiglitz-style monopolistic competition structure and a common goal of profit maximization. Third, all energy sources are exhaustible, and energy-producing sectors supply energy in order to maximize profits. Fourth, there are two types of R&D sector in the economy: one produces technologies that upgrade the quality of intermediate goods, and the other creates ESTC; they invest the same factors to maximize profits. Although the two R&D sectors employ homogeneous factors of production, they produce heterogeneous technologies due to the different output characteristics. That is why the present framework is named the framework for R&D heterogeneity-based endogenous ESTC in industry.

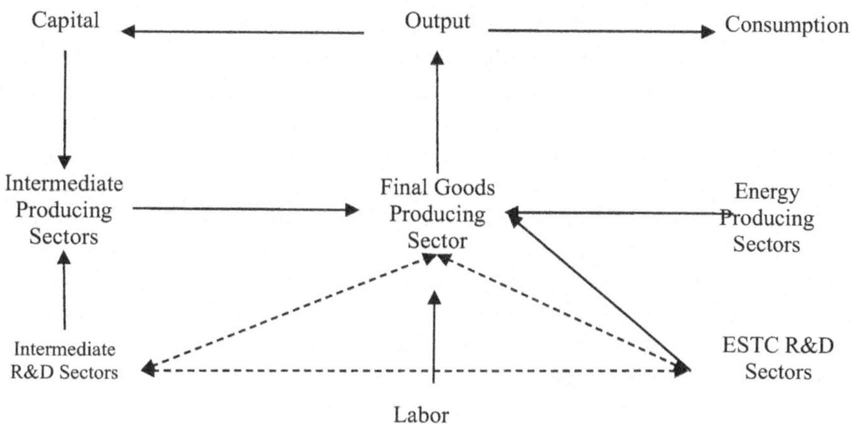

Fig. 3.2 Framework structure

The solid lines in Fig. 3.2 represent product flows, and the dashed lines mean that the sectors share the same factors of production.

3.2.1 Effect of Substitution Related to R&D Heterogeneity

In this book, I assume that different input factors in the final goods producing sector can more or less replace one another. Under this not very strict assumption, because of R&D heterogeneity, each type of R&D investment can improve energy efficiency and affect long-term economic growth, either by improving the quality of intermediate goods or by adding to effective energy input, although the two avenues operate differently, as shown in Fig. 3.3. In this sub-section, I first elaborate on the effect of substitution related to R&D heterogeneity.

As shown in Fig. 3.3, the effect of substitution related to R&D heterogeneity refers to the situation where non-energy R&D investment enters the intermediate goods R&D sectors, leading to better intermediate goods, which in turn replace part of the energy used in production, so that less energy use can be consumed to produce the same amount of output, or the same amount of energy may result in larger output. Under the more general CES production function specification (Eq. (3.8)), the effect of substitution related to R&D heterogeneity is chiefly reflected in the fact that changes in non-energy R&D investment will affect the rightmost factor in (3.8), namely, $\left(\frac{\theta_{EY}}{\gamma}\right)^{\frac{\sigma_{K,L}}{\sigma_{K,L-1}}}$. Under the specification of the CES production function, as the elasticity of factor substitution $\sigma_{K,L}$ and the share parameter γ are constant, the effect of substitution related to R&D heterogeneity means that changes in non-energy R&D investment θ_{EY} will affect energy as a share of output. The logic behind this substitution effect is that better intermediate goods will contribute to greater marginal productivity of capital, which in turn will crowd out energy in production. As the quality of capital improves, therefore, energy as a share of output will gradually decrease. If $\sigma_{K,L}$ equals 1, and the production function collapses to a Cobb–Douglas one, then (3.11) will dictate that ESTC Q_E is an increasing function of marginal productivity of capital *MPK* and a decreasing function of capital-augmenting technological change Q_K. With other conditions held fixed, if the production function remains at the isoquant curve, any increase in *MPK* or any decrease in Q_K will mean less effective capital input, resulting in more energy input; whereas any decrease in *MPK* or any increase in Q_K will means more effective capital input, which will crowd out energy input. As long as the elasticity of factor substitution is greater than or equal to 1, therefore, improvement of capital quality will crowd out energy input, thus reducing energy intensity.

3.2.2 *Effect of Increased Effective Energy Input Due to R&D Heterogeneity*

As Fig. 3.3 indicates, the effect of increased effective energy input due to R&D heterogeneity differs from the effect of factor substitution. The effect of increased

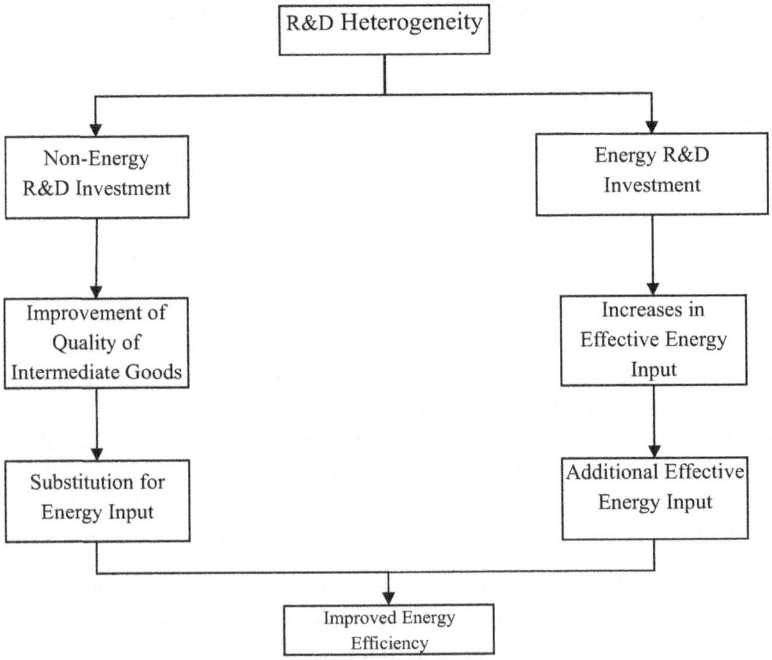

Fig. 3.3 How R&D heterogeneity improves energy efficiency

effective energy input occurs when energy R&D investment enters the ESTC R&D sectors, thus accelerating ESTC. At a set level of output of the final goods producing sector, faster ESTC makes it possible to achieve the output with less energy, resulting in energy savings. With other conditions held constant, any increase in energy R&D input, which means more effective energy input, will bring about greater aggregate output, driving energy intensity to decline. Moreover, the additional effective energy input attributable to R&D heterogeneity improves energy intensity not by crowding out energy input, rather by increasing output while keeping energy use intact. Given the neoclassical production function's assumption that factor inputs cannot form an empty set, it can be predicted through (3.8) that energy as a share of output θ_{EY} will not decrease indefinitely. When θ_{EY} falls to a certain threshold, energy's key role in production will come into play, that is, no matter how better capital becomes, the marginal productivity of capital will always be less than or equal to the marginal productivity of energy. This threshold is not necessarily a constant; as this book only deals with exhaustible resources, the marginal productivity of energy will rise with energy becoming scarcer, rendering it increasingly profitable to inject more effective energy in production. If $\sigma_{K,L} = 1$, energy as a share of output θ_{EY} is constant as shown in (3.9), and greater ESTC Q_E can ensure that energy intensity in an economy improves when energy use remains flat or increases.

3.2.3 Allocation of R&D Investment Among Different Sectors

If exogenous R&D investment in the final goods producing sector and the R&D sectors is to increase, how should this investment be allocated among the final goods producing sector, the intermediate R&D sectors and the ESTC sectors? If R&D investment is allowed to flow freely among those sectors, the most important driver of its allocation is returns generated, namely, the marginal output of those sectors. Since this book is focused on exhaustible energy, R&D investment in the ESTC sectors will generate increasing returns with the scarcity of energy intensifying. That will drive R&D investment for other sectors or additional R&D investment to enter the ESTC sectors, so as to meet the no-arbitrage condition. As long as the elasticity of substitution of factors (as between energy and capital) is greater than zero, allocation of new R&D investment will transpire like the above-described scenario, and with the elasticity of factor substitution gradually climbing, R&D investment flows to the ESTC sectors will slow down. In extreme cases where the elasticity of factor substitution is infinitely great and the production function of the final goods producing sector collapses to a linear one, with the gradual depletion of energy, R&D investment may turn into non-energy R&D investment, and will entirely flow into the intermediate R&D sectors. If the elasticity of factor substitution is zero, the production function of the final goods producing sector will collapse to a Leontief function, and scarce inputs will prove to be bottlenecks in production. As energy is being depleted, R&D investment may take the form of energy R&D investment, and all of it will flow into the ESTC sectors.

3.3 Comparison of Analytical Frameworks

In fact, the DHSS model and Dasgupta and Heal (1980) mentioned as early as 1974 the differences between the effect of factor substitution and that of increased effective energy input on energy efficiency. In a series of subsequent studies, however, researchers were bent on the development of a framework for understanding how better quality of intermediate goods (especially capital) affects energy efficiency and long-term sustainable growth, with scant attention paid to the possible effect of increased effective energy input on energy efficiency. While Hassler et al. (2012) consider in their model both types of technological change, their key assumption that capital and energy are fully complementary to each other directs their focus only on the effect of increased effective energy input at the expense of the effect of factor substitution, thus greatly limiting the applicability of endogenous ESTC theory.

3.3.1 Comparison with the DHSS Model

As its most salient feature, the DHSS model assumes that technological change is exogenous. According to Stiglitz (1974a, b), Solow (1974), and Dasgupta and Heal (1974), despite that energy supplies will dwindle over time, technological change will extricate the economy from the constraints of energy scarcity. Take (3.12) as an example:

$$y = k^a \left(Q_E \overline{E}\right)^{1-a} \tag{3.12}$$

where the symbols have the same meanings as in (3.3) except a, which means the elasticity of the marginal output of capital. Assume that the growth rate of energy is a negative constant g_E, and the steady-state growth rate of the economy near BGP will be:

$$g_Y^{ss} = g_{Q_E} + g_E \tag{3.13}$$

where g_Y^{ss} and g_{Q_E} respectively represent the steady-state growth rate and the ESTC rate. As (3.13) shows, steady-state growth is positive when and only when $g_{Q_E} > -g_E$. Given an exogenous ESTC rate, therefore, the value of g_{Q_E} must exceed a certain threshold to allow the economy to grow sustainably in the long run.

3.3.2 Comparison with Frameworks for Endogenous Embodied Technological Change

All strands of literature on the frameworks for endogenous embodied technological change assume that R&D achievements only go to the intermediate goods sectors. In other words, they set eyes solely on the effect of factor substitution on energy efficiency but choose to ignore increased effective energy input. Take Aghion and Howitt (2011) for example:

$$Y = L^{1-a}Q^{1-a}x^{a}E^{b} \tag{3.14}$$

where *Y, L, Q, x, and E* denote output, labor, quality of intermediate goods, intermediate goods input, and energy respectively. As can be seen in the equation, any increase in *Q* will crowd out part of energy input, or will boost output when energy input is kept intact. These two effects due to better intermediate goods will help improve energy intensity, though the crowding out of energy by the better quality of intermediate goods is not accompanied by more effective energy input.

3.3.3 Comparison with Frameworks for RITC in Climate Change Economics

Figure 3.4 illustrates the basic logical framework for R&D-induced technological change. As shown in Fig. 3.4 and (2.16), RITC in the main resorts to the substitution of abstract knowledge accumulation for energy and other factors used in production for sustainable growth. Examinations of the literatures on applying RITC theory to climate change and sustainable growth, conducted by researchers ranging from Goulder and Schneider (1999), Sue Wing (2001), Wang et al. (2009) to Wei Wei & He Xubo (2012, 2013), reveal that those researchers are innovative in dealing with the forms of spillovers of knowledge across different sectors, but neglect other mechanisms by which R&D investment affects ESTC.

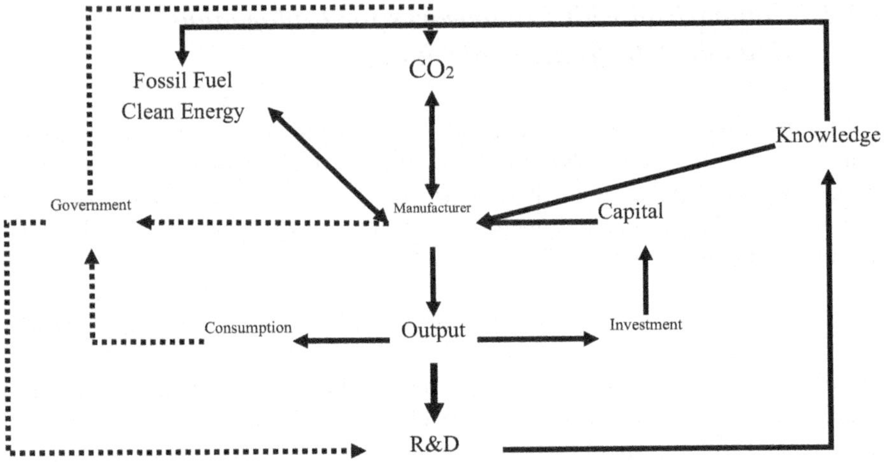

Fig. 3.4 Basic logical framework for RITC in climate change economics. Sue Wing (2001) *Source* Based on the studies by, Wei Wei & He Xubo (2012, 2013)

3.3.4 Comparison with Hassler et al.'s (2012) Framework

Hassler et al. (2012) assume that factors of production are fully complementary to one another, thus excluding from consideration the potential effect of factor substitution on energy efficiency, and adopt a Leontief production function as follows:

$$Y = \min\{Q_K K^a L^{1-a}, Q_E E\} \tag{3.15}$$

where the symbols have the same meanings as in previous chapters. Under equilibrium conditions, (3.15) morphs into $Q_K K^a L^{1-a} = Q_E E$, making steady-state ESTC equal to energy productivity.

References

Aghion, P., & Howitt, P. (2011). The economics of growth. Cambridge: MIT Press. Chinese Translation. Beijing: China Renmin University Press.

Aghion, P., & Howitt, P. (1998). *Endogenous growth theory*. MIT Press.

Binswanger H. P., & Ruttan V. W. (1978) .Induced innovation: Technology, Institution and development. Baltimore: *John Hopkins University Press*.

Chu, W., & Manhong, S. (2007a). Energy efficiency and its drivers: An empirical analysis based on DEA. *Management World, 08*, 66–76.

Chu, W., & Manhong, S. (2007b). Energy efficiency and energy productivity: A comparison of provincial data based on DEA method. *The Journal of Quantitative & Technical Economics, 09*, 110–121.

Dasgupta, P., & Heal, G. (1974). The optimal depletion of exhaustible resources. *Review of Economic Studies, 41*(5), 3–28.

Dasgupta, P. S., & Heal, G. M. (1980). *Economic theory and exhaustible resources*. Cambridge University Press.

Farrell, M. J. (1957). The measurement of productive efficiency. *Journal of Royal Statistical Society, 120*(3), 253–290.

Goulder, L. H., & Schneider, S. H. (1999). Induced technological change and the attractiveness of CO2 abatement policies. *Resource and Energy Economics, 21*(3), 211–253.

Grimaud, A., & Rouge, L. (2003). Non-renewable resources and growth with vertical innovation optimum, equilibrium and economic policies. *Journal of Environmental Economic and Management, 45*, 433–453.

Guoqiang, T. (2005). Basic analytical framework and research methods of modern economics. *Economic Research Journal, 02*, 113–125.

Hassler, J., Krusell, P., & Olovsson, C. (2012). Energy-saving technical change. NBER Working Paper No.18456.

IPCC. (2000). Emissions scenarios. Cambridge: Cambridge University Press. Available at http://www.ipcc.ch/ipccreports/sres/emission/index.php?idp=0

Jorgenson D. W., & Wilcoxen P. J. (1993). Reducing US carbon emissions: An econometric general equilibrium assessment. *Resource and Energy Economics, 15*(1), 7-25.

Lovell, C. A. K. (1995). Econometric efficiency analysis: A policy-oriented review. *European Journal of Operational Research, 80*(3), 452–461.

Newell R. G., Jaffe A. B., & Stavins R. N. (1999). The induced innovation hypothesis and energy-saving technological change. *The Quarterly Journal of Economics, 114*(3), 941-975.

Patterson, M. G. (1996). What is energy efficiency? Concepts, indictors and methodological issue. *Energy Policy, 24*(5), 377–390.

Qingwen, Y. (2012). Measurement of energy efficiency and its relationship with environmental policy tools. *Economic Studies (taiwan), 48*(2), 287–317.

Scholz, C. M., & Ziemes, G. (1999). Exhaustible resources, monopolistic competition and endogenous growth. *Environmental and Resource Economics, 13*, 169–185.

Schou, P. (2002). When environmental policy is superfluous: Growth and polluting resources. *Scandinavian Journal of Economics, 104*, 605–620.

Schumpeter. (1990). Theory of economic development. Beijing: Commercial Press.

Solow, R. M. (1974). Intergenerational equity and exhaustible resources. *Review of Economic Studies, 41*(5), 29–45.

Stiglitz, J. E. (1974a). Growth with exhaustible natural resources: Efficient and optimal growth paths. *Review of Economic Studies, 41*(5), 123–137.

Stiglitz, J. E. (1974b). Growth with exhaustible natural resources: The competitive economy. *Review of Economic Studies, 41*(5), 139–152.

Sue Wing, I. (2001). Induced technical change in computable general equilibrium models for climate-change policy analysis. PhD dissertation, Massachusetts Institute of Technology, Engineering Systems Division.

Sue Wing, I. (2003). Induced technical change and the cost of climate policy. MIT Joint Program Technical Note No. 6.

Sue, W. I. (2006). Representing induced technological change in models for climate policy. *Energy Economics, 28*(5–6), 539–562.

Tanaka, K. (2008). Assessment of energy efficiency performance measures in industry and their application for policy. *Energy Policy, 36*(8), 2887–2902.

van Zon, A., & Yetkiner, I. H. (2003). An endogenous growth model with embodied energy-saving technical change. *Resource and Energy Economics, 25*, 81–103.

Wang, K., Wang, C., & Chen, J. (2009). Analysis of the economic impact of different chinese climate policy options based on a CGE model incorporating endogenous technological change. *Energy Policy, 37*(8), 2930–2940.

Wei Chu and Shen Manhong (2007). Energy efficiency and its drivers: An empirical analysis based on DEA. *Management World, 08*, 66–76.

Wen, X., & Zhaoxi, T. (2011). Energy constraint, technological change and sustainable development—A framework of analysis based on the improvement of the quality of intermediate goods. *Economic Theory and Business Management, 01*, 87–94.

Wen, X., & Zhaoxi, T. (2012). Renewable energy, intermediate goods quality and sustainable development. Journal of World Economy, 02, 143–159.

Xiaokai, Y. (2003). *Economics: New classics and neoclassical framework*. Social Sciences Academic Press.

Xiaokai, Y., & Youguang, H. (1999). *Specialization and economic organization: A neoclassical microeconomic framework*. Economic Science Press.

Chapter 4
The Mathematical Model of R&D Heterogeneity-Based Endogenous ESTC in Industry

Based on the framework for R&D heterogeneity-based endogenous ESTC in industry introduced in Chap. 3, this chapter is to develop a mathematical model of the same theme by specifying the utility function for representative households and the production technology and economic conditions of representative manufacturers, and to compare the outcomes of the model using competitive equilibrium analysis and centrally planned equilibrium analysis.

4.1 Theoretical Arsenal

4.1.1 CES Production Function

(1) Basic Features of the CES Production Function

In early days, Arrow et al. (1961) developed the constant-elasticity-of-substitution (CES) production function incorporating two factors—capital and labor:

$$Y = \left[\gamma K^{\frac{\sigma-1}{\sigma}} + (1-\gamma)L^{\frac{\sigma-1}{\sigma}}\right]^{\frac{\sigma}{\sigma-1}} \tag{4.1}$$

where different values of σ may reduce the CES production function to specific forms as shown in Table 4.1. If $p = \frac{\sigma-1}{\sigma}$, the above equation becomes:

$$Y = \left[\gamma K^{p} + (1-\gamma)L^{p}\right]^{\frac{1}{p}} \tag{4.2}$$

X. He, *The Endogenous Energy-Saving Technological Change in China's Industrial Sector*, https://doi.org/10.1007/978-981-19-7485-4_4

Table 4.1 Specific forms of the CES production function

σ	p	Production function	Function type
0	$-\infty$	$Y = \min(K, L)$	Leontief function
1	0	$Y = K^{\gamma} L^{1-\gamma}$	Cobb–Douglas function
∞	1	$Y = \gamma K + (1-\gamma)L$	Linear function

Equation (4.2) can be considered a specific form of the general mean of order p (de La Grandville, 2009), in which the value of Y is an increasing function of order p.

(2) Nested CES Production Function

After Arrow et al. (1961), Uzawa (1961) expanded the CES production function to include multiple production factors, and Sato (1967) introduced the basic concept of the nested CES function. In a two-factor CES function, both factors enter the CES function simultaneously, whereas in a multi-factor nested CES function, what combinations of factors are admitted into the production function is especially important, because different combinations not only lead to different calculation results, but have different economic implications.

The CES production function appears in only one form when it encompasses only two factors, but takes at least four forms when more than two factors (e.g., capital, labor, and energy) are involved. With empirical pillars of the industrial production function in China yet to be established, it is hard to choose the "right" form of the CES function to describe the technology employed by the final goods producing sector. Below, I use three factors of production, namely, capital (K), labor (L) and energy (E), for example, to illustrate the specification of the function. When more than two factors are present in the CES function, the first two may unite to make intermediate goods, before producing final goods with the help of the third. Alternatively, the three factors may play their parts in the function at the same time to produce final goods. A three-factor CES production function, therefore, has at least four nested forms: (KE)L, (KL)E, (LE)K, and (KLE). In particular, (KL)E means that capital and labor work together to yield intermediate goods Z_{KL}, then take on energy to produce final goods Y, as shown in Eqs. (4.3) and (4.4):

$$Y = \left[\alpha_1 (Q_E E)^{\frac{\sigma_{KL,E}-1}{\sigma_{KL,E}}} + (1-\alpha_1)(Z_{KL})^{\frac{\sigma_{KL,E}-1}{\sigma_{KL,E}}}\right]^{\frac{\sigma_{KL,E}}{\sigma_{KL,E}-1}} \tag{4.3}$$

$$Z_{KL} = \left[\beta_1 (Q_K K)^{\frac{\sigma_{K,L}-1}{\sigma_{K,L}}} + (1-\beta_1)(Q_L L)^{\frac{\sigma_{K,L}-1}{\sigma_{K,L}}}\right]^{\frac{\sigma_{K,L}}{\sigma_{K,L}-1}} \tag{4.4}$$

Symbols in Eqs. (4.3) and (4.4) have the same meanings as those in Chap. 3, and the subscript to the elasticity of substitution σ points to the elasticity of substitution between different factors. The other three forms of the CES function are similar to Eqs. (4.3) and (4.4), and can be expressed as follows respectively:

(LE)K:

$$Y = \left[\alpha_2(Q_K K)^{\frac{\sigma_{LE,K}-1}{\sigma_{LE,K}}} + (1-\alpha_2)(Z_{LE})^{\frac{\sigma_{LE,K}-1}{\sigma_{LE,K}}}\right]^{\frac{\sigma_{LE,K}}{\sigma_{LE,K}-1}} \tag{4.5}$$

$$Z_{LE} = \left[\beta_2(Q_E E)^{\frac{\sigma_{L,E}-1}{\sigma_{L,E}}} + (1-\beta_2)(Q_L L)^{\frac{\sigma_{L,E}-1}{\sigma_{L,E}}}\right]^{\frac{\sigma_{L,E}}{\sigma_{L,E}-1}} \tag{4.6}$$

(KE)L:

$$Y = \left[\alpha_3(Q_L L)^{\frac{\sigma_{KE,L}-1}{\sigma_{KE,L}}} + (1-\alpha_3)(Z_{KE})^{\frac{\sigma_{KE,L}-1}{\sigma_{KE,L}}}\right]^{\frac{\sigma_{KE,L}}{\sigma_{KE,L}-1}} \tag{4.7}$$

$$Z_{KE} = \left[\beta_3(Q_E E)^{\frac{\sigma_{K,E}-1}{\sigma_{K,E}}} + (1-\beta_3)(Q_K K)^{\frac{\sigma_{K,E}-1}{\sigma_{K,E}}}\right]^{\frac{\sigma_{K,E}}{\sigma_{K,E}-1}} \tag{4.8}$$

(KLE):

$$Y = \left[\alpha_4(Q_L L)^{\frac{\sigma_{KLE}-1}{\sigma_{KLE}}} + \beta_4(Q_E E)^{\frac{\sigma_{KLE}-1}{\sigma_{KLE}}} + (1-\alpha_3-\beta_4)(Q_K K)^{\frac{\sigma_{KLE}-1}{\sigma_{KLE}}}\right]^{\frac{\sigma_{KLE}}{\sigma_{KLE}-1}} \tag{4.9}$$

As Eqs. (4.3), (4.4), (4.5), (4.6), (4.7), (4.8) and (4.9) indicate, with no empirical support, it is difficult to determine the form of a three-factor CES function.

To establish the production function for the final goods producing sector and explore the different avenues through which R&D heterogeneity works, a viable solution is to simplify the production function by assuming from the outset that the elasticity of substitution between all factors is always 1, that is, $\sigma_{KLE} = \sigma_{K,E} = \sigma_{KE,L} = \sigma_{L,E} = \sigma_{LE,K} = \sigma_{K,L} = \sigma_{KL,E} \equiv 1$. Under this assumption, each of (KE)L, (KL)E, (LE)K, and (KLE) collapses to being Cobb–Douglas. This assumption is based on three reasons. First, as indicated in Eqs. (3.7) and (3.10), the elasticity of substitution between production factors is positive under either the CES production function or the Cobb–Douglas production function, so that both production functions feature the effect of factor substitution and that of increased effective energy input. Second, the assumption enables an analytical solution to a steady-state economic system, which is hard to come by under the CES function (de La Grandville, 2009). Third, as demonstrated in Sect. 2.2, both an elasticity of factor substitution equal to or greater than 1 and ESTC can elevate energy efficiency. Table 4.2 summarizes the main conclusions drawn from the literature listed in Sect. 2.2.

The goal of this book should not be compromised, therefore, if I set aside other possible values of substitution elasticity but focus on how ESTC can be endogenous and through what path it plays its role under the Cobb–Douglas production function. In Sect. 4.4 of this chapter, I will loosen the assumption of the Cobb–Douglas

Table 4.2 How Substitution elasticity and technological change are related to sustainable growth

	Elasticity of substitution less than 1	Elasticity of substitution equal to 1	Elasticity of substitution greater than 1
Absence of technological change	No growth	Conditional growth	Growth
Presence of technological change	Conditional growth	Conditional growth	Growth

Source Based on the relevant literature listed in Sect. 2.2
Note The condition for growth that requires no technological change and a substitution elasticity of 1 means that the elasticity of the marginal output of capital is greater than that of energy, as elaborated by Solow (1974). The combination of technological change and a substitution elasticity less than or equal to 1 as a condition for growth means that ESTC is present and the rate of technological change is greater than the depreciation rate, as detailed in Dasgupta and Heal (1980) and Stiglitz (1974a, b). When substitution elasticity is greater than 1, energy is no longer indispensable in the production function, thus the lack of it will not hinder long-term economic growth

production function to concentrate on how substitution elasticity impacts ESTC and sustainable growth.

In sum, in order to investigate the two ways in which R&D heterogeneity works, a practical modeling method is first to skip the effects of other possible values of substitution elasticity on the path to balanced growth, but consider how R&D heterogeneity affects ESTC when factors of production are substitutable and ESTC exists under the Cobb–Douglas production function. The next step is to relax the assumption of the Cobb–Douglas production function, and reasonably simplify the model to highlight the effects of capital and changes in energy substitution elasticity on the allocation of R&D investment and the path to balanced growth. In Chap. 5, I will loosen the above assumption about substitution elasticity, and estimate all possible forms of a three-factor CES function in Chinese industry.

4.2 Competitive Equilibrium Analysis

4.2.1 Description of Economic Conditions

(1) Production Technology

(1) Final Goods Producing Sector

Suppose there is only one final goods producing sector in the economy, running on a production function as shown in Eq. (4.10), under which labor L_Y, intermediate goods $x(i)$ and energy E are employed to produce final goods in period t; Q_K and Q_E respectively represent the quality of intermediate goods and ESTC; labor-augmenting technological change Q_L is constant at 1; intermediate goods are treated by the same method as adopted by Dixit and Stiglitz (1977), and the types of intermediate goods are a continuum normalized as 1; and labor, intermediate goods and energy jointly enter the Cobb–Douglas production function of the final goods producing sector to organize production. Then:

$$Y_t = L_{Yt}^{a_1}\left[\int_0^1 Q_{Kt}(i)x_t(i)^{a_2}\mathrm{d}i\right](Q_{Et}E_t)^{a_3} \tag{4.10}$$

where $a_1, a_2, a_3 > 0$ and $a_1 + a_2 + a_3 = 1$, indicating that returns to scale of the final goods producing sector remains constant. Assume that the final goods are only used for consumption and investment, and the capital depreciation rate is zero. Then:

$$Y_t = C_t + \dot{K}_t \tag{4.11}$$

where $\dot{K}_t$ is the derivative of capital stock with respect to time, and is equal to the investment in period t when the capital depreciation rate is zero. Unless otherwise noted, $*$ denotes the derivative of a variable with respect to time.

(2) Intermediate Goods Producing Sectors

Assume that the output $x(i)$ of an intermediate goods sector is a linear function of capital input $K(i)$ and more capital is needed to produce high-quality intermediate goods. Then:

$$x_t(i) = \frac{K_t(i)}{Q_{Kt}(i)} \tag{4.12}$$

According to Eq. (4.10), the total stock of capital K_t in period t satisfies:

$$K_t = \int_0^1 Q_{Kt}(i)x_t(i)\mathrm{d}i \tag{4.13}$$

(3) Energy Producing Sectors

Assume the cost of extracting energy is zero. I apply Hotelling's (1931) method, and let the dynamics of energy extraction satisfy Eq. (4.14):

$$\dot{S}_t = -E_t \tag{4.14}$$

where S_t 和 E_t respectively denote the stock of energy and the amount of energy extracted during period t, and $\dot{S}_t$, $E_t > 0$.

(4) Heterogeneous R&D Sectors

One of my core assumptions in this book is that there are two types of R&D sector in the economy, one producing technology Q_K that helps make better intermediate goods, and the other ESTC. The outputs of both sectors fit a Poisson distribution with the parameter λ, and it requires only the input of homogeneous labor to produce these two types of technology. In a given period of time dt, the two R&D sectors produce $\lambda L_{Kt}\mathrm{d}t$ and $\lambda L_{Et}\mathrm{d}t$ respectively, in which L_{Kt} and L_{Et} respectively denote the level of labor required to improve the quality of intermediate goods and enhance ESTC. Assume that the incremental outputs of the two R&D sectors are η_K and η_E, that is $Q_{K,t+1} = (1+\eta_K)Q_{Kt}$ and $Q_{E,t+1} = (1+\eta_E)Q_{Et}$, then the knowledge production functions will be:

$$\dot{Q}_{Kt} = \lambda\eta_K L_{Kt} Q_{Kt} \tag{4.15}$$

$$\dot{Q}_{Et} = \lambda\eta_E L_{Et} Q_{Et} \tag{4.16}$$

(2) Consumer Utility

Assume that the utility function of representative households is represented by the Constant Relative Risk Aversion (CRRA) utility function. That is, the instantaneous utility of representative households is only related to its consumption level C and the coefficient of relative risk aversion θ. As the reciprocal of the elasticity of intertemporal substitution, the value of θ reflects representative households' attitude towards how to shift consumption over time. In an infinite timeframe, the discounted lifetime utility of representative households is:

$$U = \int_0^\infty \frac{C_t^{1-\theta} - 1}{1-\theta} e^{-\rho t}\mathrm{d}t \tag{4.17}$$

where ρ is the subjective discount rate of representative households: a larger value of ρ means that representative households favor current consumption, whereas a smaller value of ρ indicates their preference for future consumption. To simplify the analysis, I assume that the coefficient of relative risk aversion is 1. Then Eq. (4.17) becomes:

$$U = \int_0^\infty e^{-\rho t} \ln C_t \mathrm{d}t \tag{4.18}$$

4.2.2 *Balanced Growth Path in Competitive Equilibrium*

(1) Final Goods Producing Sector

To simplify the derivation process, I omit subscript t denoting time where there is no danger of confusion. As implied by the economic conditions described above, it is optimal for the final goods producing sector that representative manufacturers organize production after they hire labor and purchase energy, intermediate goods, and ESTC in pursuit of maximum profit π_Y, as illustrated in Eq. (4.19):

$$\max_{L_Y, x(i), E, Q_E} \pi_Y = L_Y^{a_1}\left[\int_0^1 Q_K(i)x(i)^{a_2}\mathrm{d}i\right](Q_E E)^{a_3} - w_Y L_Y - p_E E - \int_0^1 p_{x_i} x(i)\mathrm{d}i - p_{Q_E} Q_E \tag{4.19}$$

where, if the price of a final good is normalized to be 1, the prices of other inputs (i.e., labor, energy, capital, and ESTC) can be regarded as prices relative to that final good. Use Eq. (4.19) to calculate the partial derivatives of labor, energy, intermediate goods and ESTC, and the optimal behavior of the final goods producing sector is captured by Eqs. (4.20), (4.21), (4.22) and (4.23) below:

$$\frac{\partial \pi_Y}{\partial L_Y} = \frac{a_1 Y}{L_Y} - w_Y = 0 \tag{4.20}$$

$$\frac{\partial \pi_Y}{\partial E} = \frac{a_3 Y}{E} - p_E = 0 \tag{4.21}$$

$$\frac{\partial \pi_Y}{\partial x(i)} = \frac{a_2 Y}{x(i)} - p_{x_i} = 0 \tag{4.22}$$

$$\frac{\partial \pi_Y}{\partial Q_E} = \frac{a_3 Y}{Q_E} - p_{Q_E} = 0 \tag{4.23}$$

As shown by Eqs. (4.20), (4.21), (4.22) and (4.23), the final goods producing sector will keep hiring or purchasing labor, energy, intermediate goods and ESTC until the marginal outputs of those production factors equal their marginal costs, namely, their relative prices. Substituting Eq. (4.10) into (4.22) will yield an expression for the price of type i intermediate good:

$$p_{xi} = a_2 L_Y^{a_1} Q_K(i)x(i)^{a_2-1}(Q_E E)^{a_3} \tag{4.24}$$

(2) Intermediate Goods Producing Sectors

Assume that the market interest rate is a constant r. As implied by the assumption that the output of an intermediate goods sector is a linear function of capital, optimally an intermediate goods sector will produce $x(i)$ to maximize its profit, as shown by Eq. (4.25):

$$\max_{x(i)} \pi_{x(i)} = p_{x_i} x(i) - rQ_K(i)x(i) \tag{4.25}$$

Substituting Eq. (4.24) into (4.25) and solving for the first-order conditions of Eq. (4.25) will lead to (4.26):

$$p_{x_i} = \frac{rQ_K(i)}{a_2} \tag{4.26}$$

Substituting Eq. (4.24) into (4.26), we get Eq. (4.27) by dropping p_{x_i}:

$$x(i) = \left[\frac{a_2^2 L_Y^{a_1} (Q_E E)^{a_3}}{r}\right]^{\frac{1}{1-a_2}} \tag{4.27}$$

As the terms in brackets on the right-hand side of Eq. (4.27) are unrelated to type i intermediate good, the optimal decision of an intermediate goods sector in equilibrium is to make the same amount of each type of intermediate goods, as shown by Eq. (4.28):

$$x(i) = x = \frac{K}{Q_K} \tag{4.28}$$

where $K = \int_0^1 Q_K(i)x(i)\mathrm{d}i$, and $Q_K = \int_0^1 Q_K(i)\mathrm{d}i$, and Q_K can be taken as the average quality of the intermediate goods. Put Eq. (4.28) and the expressions for K and Q_K back to Eq. (4.10) of the final goods producing sector, then we get another expression of the production function of the final goods producing sector:

$$Y = L_Y^{a_1} Q_K^{1-a_2} K^{a_2} (Q_E E)^{a_3} \tag{4.29}$$

(3) Energy Producing Sectors

Since the cost of energy extraction is zero, the optimal decision of an energy producing sector in a fully competitive market must satisfy:

$$\begin{aligned} &\max_{E_t} \pi_E = \int_t^{\infty} p_E E_\tau e^{-\int_t^\tau r_u \mathrm{d}\tau} \\ &s.t. \dot{S} = -E \end{aligned} \tag{4.30}$$

where p_E is the energy price. As under Eq. (4.30) energy producing sectors are bent on maximizing the sum of the discounted values of all revenues generated with a limited stock of energy, the current value Hamiltonian equation for energy producing sectors may be expressed as:

$$H_S = p_E E + \mu_S(-E) \tag{4.31}$$

where E is a control variable; S is a state variable; and μ_S is a co-state variable, which can be regarded as the shadow price of the stock of energy S. From Eq. (4.30) flow the following first-order conditions:

$$\frac{\partial H_E}{\partial E} = p_E - \mu_S = 0 \tag{4.32}$$

$$\frac{\partial H_E}{\partial S} = \rho\mu_S - \dot{\mu}_S = 0 \tag{4.33}$$

The transversality condition is:

$$\lim_{t\to\infty} \mu_S S_t e^{-\rho t} = 0 \tag{4.34}$$

which guarantees that energy producing sectors will exploit all energy reserves to maximize the sum of the discounted values of all their revenues. Using the same symbols as in Chap. 3, and letting g_* be the growth rate, we get from Eqs. (4.32) and (4.33):

$$g_{p_E} = r \tag{4.35}$$

Equation (4.35) is also called the Hotelling Rule (Hotelling, 1931), which states that in equilibrium the growth rate of energy price is equal to the market rate of interest.

(4) Heterogeneous R&D Sectors

Since R&D investment in the heterogeneous R&D sectors is homogeneous labor, profits for those sectors should satisfy:

$$\pi_{Q_K} = p_{Q_K}\dot{Q}_K - w_K L_K \tag{4.36}$$

$$\pi_{Q_E} = p_{Q_E}\dot{Q}_E - w_E L_E \tag{4.37}$$

If the heterogeneous R&D sectors all run after maximum profits, their optimal decisions should conform to:

$$\max_{L_K} \pi_{Q_K} = p_{Q_K}\lambda\eta_K L_K Q_K - w_K L_K \tag{4.38}$$

$$\max_{L_E} \pi_{Q_E} = p_{Q_E}\lambda\eta_E L_E Q_E - w_E L_E \tag{4.39}$$

whose first-order conditions are:

$$w_K = p_{Q_K}\lambda\eta_K Q_K \tag{4.40}$$

$$w_E = p_{Q_E}\lambda\eta_E Q_E \tag{4.41}$$

Using the Romer model (1990), I assume, for the sake of simplicity, the aggregate level of labor supply is an exogenous constant $\overline{L}$. As the labor force employed by the final goods producing sector is homogeneous with that used by the heterogeneous

R&D sectors, namely, $L_{Yt} + L_{Kt} + L_{Et} = \overline{L}$, assuming that labor can flow freely between the final goods producing sector and the two heterogeneous R&D sectors, we can deduct from the no-arbitrage condition that the final goods producing sector, the intermediate goods quality R&D sectors and the ESTC R&D sectors in equilibrium provide equal wage rates.

(5) Representative Households

Under standard neoclassical growth theory, if the utility function of representative households is of CRRA form, then Eq. (4.42) is indispensable in representative households' decision to maximize utility (Eq. (4.17)):

$$g_C = r - \rho \tag{4.42}$$

Equation (4.42) is also known as the turnpike theorem, which states that the growth rate of consumption is related to the market rate of interest and representative households' subjective discount rate.

(6) Optimal Path to Economic Equilibrium in Competitive Equilibrium

The paths to equilibrium in various sectors converge to form the path to equilibrium of the whole system, as illustrated by Proposition 4.1 below:

Proposition 4.1 Under the economic conditions described by Eqs. (4.10), (4.11), (4.12), (4.13), (4.14), (4.15), (4.16) and (4.18), if every economic actor pursues its optimal decision, that is, the final goods producing sector, the intermediate goods producing sectors, the energy producing sectors, and the heterogeneous R&D sectors all seek to maximize profits, and representative households strive to maximize utility, then the equilibrium growth rates of the variables in steady state are:

$$g_Y^{ss} = g_C^{ss} = \frac{(1-a_2)\lambda\eta_K L_K^{ss} + a_3\lambda\eta_E L_E^{ss} - a_3\rho}{a_1 + a_3} \tag{4.43}$$

$$g_{Q_K}^{ss} = \lambda\eta_K L_K^{ss} \tag{4.44}$$

$$g_{Q_E}^{ss} = \lambda\eta_E L_E^{ss} \tag{4.45}$$

$$g_E^{ss} = -\rho \tag{4.46}$$

where L_Y^{ss} and L_K^{ss} satisfy:

$$L_Y^{ss} = \frac{a_1}{a_3\lambda\eta_E} \tag{4.47}$$

$$L_K^{ss} = \frac{(a_1+a_3)(1-a_2)a_2 - a_1(a_1+a_2)\rho}{a_1(a_1+a_3)\lambda\eta_K} \tag{4.48}$$

Superscript ss in the above equations means that the relevant variables are in steady state.

Since the steady-state energy intensity growth rate $g_{In}^{ss} = g_E^{ss} - g_Y^{ss}$, it will decline in equilibrium as a consequence of either better intermediate goods or more effective energy input, as shown by Eqs. (4.43) and (4.46). That is to say, ESTC is a requisite for steady-state energy intensity improvements. In addition, as shown by Eq. (4.43), increases in energy R&D investment and non-energy R&D investment in equilibrium are all effective at lowering energy intensity and boosting economic growth.

4.2.3 Comparative Statics Analysis

Comparative statics is chiefly used to evaluate how changes in the parameters of a model affect the major variables in an economy and long-term growth. Holding other parameters of the model fixed, we can calculate the partial derivatives of all the variables in steady state with respect to one of those parameters to see whether changes in the parameter have positive or negative effects on the variables in equilibrium. Major parameters of the model include the subjective discount rate of the representative household ρ, the aggregate level of labor supply $\overline{L}$, and the incremental outputs of the heterogeneous R&D sectors η_K and η_E. Substitute Eqs. (4.41) and (4.42) into (4.37), (4.38) and (4.39) and replace the variables near their points of equilibrium with exogenous parameters to obtain:

$$\begin{aligned} r^{ss} = & \frac{a_2(a_1+a_3)}{a_1} - \frac{a_2}{a_1+a_3}\rho + \frac{a_3}{a_1+a_3}\lambda\eta_E\overline{L} \\ & - \frac{a_1}{a_1+a_3} - \frac{a_2a_3}{a_1}\frac{\eta_E}{\eta_K} + \frac{a_3(a_1+a_2)}{(a_1+a_3)^2}\frac{\rho\eta_E}{\eta_K} \end{aligned} \tag{4.49}$$

$$\begin{aligned} g_Y^{ss} = & \frac{a_2(a_1+a_3)}{a_1} - \frac{1}{a_1+a_3}\rho + \frac{a_3}{a_1+a_3}\lambda\eta_E\overline{L} \\ & - \frac{a_1}{a_1+a_3} - \frac{a_2a_3}{a_1}\frac{\eta_E}{\eta_K} + \frac{a_3(a_1+a_2)}{(a_1+a_3)^2}\frac{\rho\eta_E}{\eta_K} \end{aligned} \tag{4.50}$$

$$g_E^{ss} = \rho \tag{4.51}$$

$$L_K^{ss} = \frac{(a_1+a_3)(1-a_2)a_2 - a_1(a_1+a_2)\rho}{a_1(a_1+a_3)\lambda\eta_K} \tag{4.52}$$

$$L_Y^{ss} = \frac{a_1}{a_3\lambda\eta_E} \tag{4.53}$$

$$g_{Q_K}^{ss} = \frac{a_2(a_1+a_3)}{a_1} - \frac{a_1+a_2}{a_1+a_3}\rho \tag{4.54}$$

Table 4.3 Comparative statics analysis of optimal economic equilibrium in competitive equilibrium

	$* = \rho$	$* = \overline{L}$	$* = \eta_K$	$* = \eta_E$
$\frac{\partial r^{ss}}{\partial *}$	>0 if $\eta_E > \frac{a_2(a_1+a_3)}{a_1+a_2}\eta_K$	>0	>0 if $\rho < \frac{a_2(a_1+a_3)^2}{a_1(a_1+a_2)}$	>0 if $\eta_K < \frac{a_2(a_1+a_3)}{a_1 \lambda \overline{L}}$
$\frac{\partial g_Y^{ss}}{\partial *}$	>0 if $\eta_E > \frac{(a_1+a_3)(a_1-a_2+a_3)}{a_1+a_2}\eta_K$	>0	>0 if $\rho < \frac{a_2(a_1+a_3)^2}{a_1(a_1+a_2)}$	>0 if $\eta_K < \frac{a_2(a_1+a_3)}{a_1 \lambda \overline{L}}$
$\frac{\partial g_E^{ss}}{\partial *}$	<0	=0	=0	=0
$\frac{\partial L_Y^{ss}}{\partial *}$	=0	=0	=0	<0
$\frac{\partial L_K^{ss}}{\partial *}$	<0	=0	<0	=0
$\frac{\partial L_E^{ss}}{\partial *}$	>0	>0	>0	>0
$\frac{\partial g_{Q_K}^{ss}}{\partial *}$	<0	=0	=0	=0
$\frac{\partial g_{Q_E}^{ss}}{\partial *}$	>0	>0	>0 if $\rho < \frac{a_2(a_1+a_3)^2}{a_1(a_1+a_2)}$	>0 if $\eta_K < \frac{a_2(a_1+a_3)}{a_1 \lambda \overline{L}}$

$$g_{Q_E}^{ss} = \lambda \eta_E \overline{L} - \frac{a_1}{a_3} - \frac{a_2(a_1+a_3)}{a_1}\frac{\eta_E}{\eta_K} + \frac{a_1+a_2}{a_1+a_3}\frac{\rho \eta_E}{\eta_K} \tag{4.55}$$

The partial derivatives of ρ, $\overline{L}$, and η_K and η_E calculated using Eqs. (4.49), (4.50), (4.51), (4.52), (4.53), (4.54) and (4.55) are summarized in Table 4.3.

As revealed by Table 4.3, under the economic conditions described by Eqs. (4.10), (4.11), (4.12), (4.13), (4.14), (4.15), (4.16) and (4.18), the following inferences are valid in competitive equilibrium:

Inference C-1 If $\eta_E > \frac{a_2(a_1+a_3)}{a_1+a_2}\eta_K$, then the steady-state market rate of interest r^{ss} is an increasing function of the subjective discount rate ρ; r^{ss} is an increasing function of the aggregate level of labor supply $\overline{L}$; If $\rho < \frac{a_2(a_1+a_3)^2}{a_1(a_1+a_2)}$, then r^{ss} is an increasing function of η_K; if $\eta_K < \frac{a_2(a_1+a_3)}{a_1 \lambda \overline{L}}$, then r^{ss} is an increasing function of η_E.

Inference C-2 If $\eta_E > \frac{(a_1+a_3)(a_1-a_2+a_3)}{a_1+a_2}\eta_K$, then the steady-state growth rate g_Y^{ss} is an increasing function of the subjective discount rate ρ; g_Y^{ss} is an increasing function of $\overline{L}$; if $\rho < \frac{a_2(a_1+a_3)^2}{a_1(a_1+a_2)}$, then g_Y^{ss} is an increasing function of η_K; if $\eta_K < \frac{a_2(a_1+a_3)}{a_1 \lambda \overline{L}}$, then g_Y^{ss} is an increasing function of η_E.

Inference C-3 The steady-state growth rate of energy consumption g_E^{ss} is only a decreasing function of ρ.

Inference C-4 Labor employed by the final goods producing sector in steady-state L_Y^{ss} is only a decreasing function of η_E.

Inference C-5 Steady-state non-energy R&D investment in the intermediate goods producing sectors in steady state L_K^{ss} is an decreasing function of ρ and η_K.

Inference C-6 Steady-state energy R&D investment in the ESTC R&D sector L_E^{ss} is an increasing function of ρ, $\overline{L}$, η_K and η_E.

Inference C-7 The steady-state rate of quality improvement of intermediate goods $g_{Q_K}^{ss}$ is only a decreasing function of ρ.

Inference C-8 The steady-state rate of ESTC $g_{Q_E}^{ss}$ is an increasing function of ρ and $\overline{L}$; if $\rho < \frac{a_2(a_1+a_3)^2}{a_1(a_1+a_2)}$, then $g_{Q_E}^{ss}$ is an increasing function of η_K; if $\eta_K < \frac{a_2(a_1+a_3)}{a_1\lambda\overline{L}}$, then $g_{Q_E}^{ss}$ is an increasing function of η_E.

4.2.4 *Numerical Simulation Analysis*

The results of numerical simulation are shown in Table 4.4. The reference parameters in Table 4.4 (the first line) mainly originate from Wen and Zhaoxi (2012)[1]; g_{In}^{ss} in the last column is the equilibrium rate of change in energy intensity, that is, $g_{In}^{ss} = g_E^{ss} - g_Y^{ss}$. Figure 4.1 shows the equilibrium rate of ESTC $g_{Q_E}^{ss}$ and g_{In}^{ss} at given different parameters.

Table 4.4 Numerical simulation of optimal economic equilibrium in competitive equilibrium

a_1	a_2	a_3	ρ	$\overline{L}$	λ	η_K	η_E	L_Y^{ss}	L_K^{ss}	L_E^{ss}	g_Y^{ss}	$g_{Q_K}^{ss}$	$g_{Q_E}^{ss}$	g_{In}^{ss}
0.4	0.4	0.2	0.02	20	1	0.2	0.2	10	2.87	7.13	1.04	0.57	1.43	−1.06
0.4	0.4	0.2	0.04	20	1	0.2	0.2	10	2.73	7.27	1.02	0.55	1.45	−1.06
0.4	0.4	0.2	0.01	20	1	0.2	0.2	10	2.93	7.07	1.05	0.59	1.41	−1.06
0.4	0.4	0.2	0.02	30	1	0.2	0.2	10	2.87	17.13	1.71	0.57	3.43	−1.73
0.4	0.4	0.2	0.02	20	2	0.2	0.2	5	1.43	13.57	2.38	0.57	5.43	−2.40
0.4	0.4	0.2	0.02	20	0.5	0.2	0.2	20	5.73	−5.73	0.38	0.57	−0.57	−0.40
0.4	0.4	0.2	0.02	15	1	0.2	0.2	10	2.87	2.13	0.71	0.57	0.43	−0.73
0.4	0.4	0.2	0.02	20	1	0.4	0.2	10	1.43	8.57	1.14	0.57	1.71	−1.16
0.4	0.4	0.2	0.02	20	1	0.2	0.4	5	2.87	12.13	2.18	0.57	4.85	−2.20
0.4	0.4	0.2	0.02	20	1	0.4	0.4	5	1.43	13.57	2.38	0.57	5.43	−2.40
0.6	0.2	0.2	0.02	20	1	0.2	0.2	15	1.23	3.77	0.43	0.25	0.75	−0.45
0.2	0.6	0.2	0.02	20	1	0.2	0.2	5	5.8	9.2	2.07	1.16	1.84	−2.09
0.2	0.2	0.6	0.02	20	1	0.2	0.2	1.67	3.95	14.38	2.93	0.79	2.88	−2.95
0.4	0.3	0.3	0.02	20	1	0.2	0.2	6.67	2.53	10.81	1.42	0.51	2.16	−1.44

[1] As Wen and Zhaoxi (2012) do not mention the ESTC R&D sector, I assume that the reference parameters of this sector are identical with those of the intermediate R&D sectors.

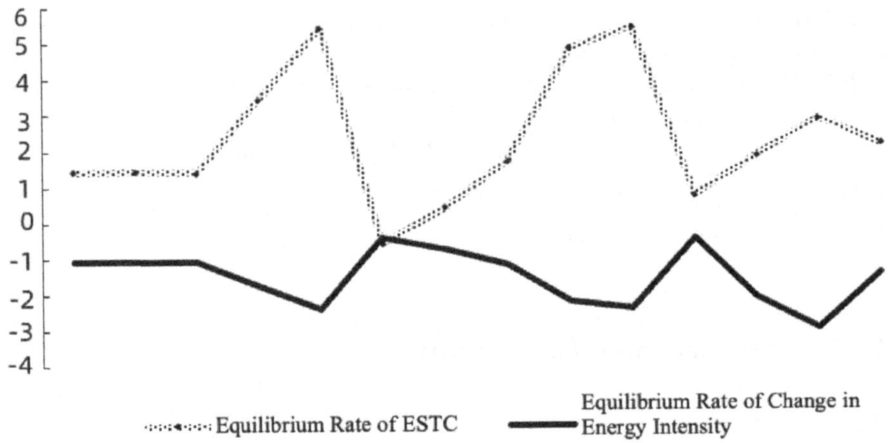

Fig. 4.1 Equilibrium rate of ESTC and equilibrium rate of change in energy intensity in the economy in competitive equilibrium

4.3 Centrally Planned Equilibrium Analysis

4.3.1 Description of Economic Conditions

Unlike profit-maximizing firms and utility-maximizing consumers in competitive equilibrium, central planners come to grips with the question of how, given the production technology Eq. (4.29) and dynamic Eqs. (4.11), (4.14), (4.15), (4.16), (4.17), (4.18), (4.19), (4.20), (4.21), (4.22), (4.23), (4.24), (4.25) and (4.26), to allocate resources to maximize the well-being of representative households (Eq. (4.17)). This question can be expressed as one of optimal growth by the following system of equations:

$$
\begin{aligned}
&\max_{C_t} U = \int_0^{\infty} e^{-\rho t} \ln C_t \mathrm{d}t \\
&s.t.\ Y_t = L_{Y_t}^{a_1} Q_{K_t}^{1-a_2} K_t^{a_2} \left(Q_{E_t} E_t\right)^{a_3} \\
&Y_t = C_t + \dot{K}_t \\
&\dot{S}_t = -E_t \\
&\dot{Q}_{Kt} = \lambda \eta_K L_{Kt} Q_{Kt} \\
&\dot{Q}_{Et} = \lambda \eta_E L_{Et} Q_{Et} \\
&L_{Yt} + L_{Kt} + L_{Et} = \overline{L} \\
&\text{where } C_t,\ K_t,\ S_t,\ E_t,\ L_{Y_t},\ L_{K_t},\ L_{E_t},\ Q_{K_t},\ Q_{E_t} \geq 0,\ K_0,\ S_0,\ Q_{K_0},\ Q_{E_0}.
\end{aligned}
\tag{4.56}
$$

4.3.2 Balanced Growth Path in Centrally Planned Equilibrium

As in Sect. 4.2.2 above, the following analysis leaves out the time subscript t for simplicity purposes where no confusion will be caused. Equation (4.56) that represents the question of optimal growth faced by central planners may be rewritten as a current-value Hamiltonian function:

$$
\begin{aligned}
H = \ln C + \mu_1\left[L_Y^{a_1} Q_K^{1-a_2} K^{a_2} (Q_E E)^{a_3} - C\right] \\
+ \mu_2[-E] + \mu_3\left[\lambda\eta_K L_K Q_K\right] + \mu_4\left[\lambda\eta_E\left(\bar{L} - L_Y - L_K\right)Q_E\right]
\end{aligned} \tag{4.57}
$$

where C, E, L_Y and L_K are control variables; K, S, Q_K and Q_E are state variables; μ_1 is a Lagrange multiplier; and μ_2, μ_3 and μ_4 are co-state variables. Given $L_Y + L_K + L_E = \bar{L}$, Eq. (4.57) may be rewritten as the production function for the ESTC R&D sector. Solve for the first-order condition for Eq. (4.57) to get:

$$\frac{\partial H}{\partial C} = \frac{1}{C} - \mu_1 = 0 \tag{4.58}$$

$$\frac{\partial H}{\partial E} = \mu_1 \frac{a_3 Y}{E} - \mu_2 = 0 \tag{4.59}$$

$$\frac{\partial H}{\partial L_Y} = \mu_1 \frac{a_1 Y}{L_Y} - \mu_4 \lambda \eta_E Q_E = 0 \tag{4.60}$$

$$\frac{\partial H}{\partial L_K} = \mu_3 \lambda \eta_K Q_K - \mu_4 \lambda \eta_E Q_E = 0 \tag{4.61}$$

$$\frac{\partial H}{\partial K} = \mu_1 \frac{a_2 Y}{K} = \rho\mu_1 - \dot{\mu}_1 \tag{4.62}$$

$$\frac{\partial H}{\partial S} = \rho\mu_2 - \dot{\mu}_2 = 0 \tag{4.63}$$

$$\frac{\partial H}{\partial Q_K} = \mu_1 \frac{(1-a_2)Y}{Q_K} + \mu_3 \lambda \eta_K L_K = \rho\mu_3 - \dot{\mu}_3 \tag{4.64}$$

$$\frac{\partial H}{\partial Q_E} = \mu_1 \frac{a_3 Y}{Q_E} + \mu_4 \lambda \eta_E \left(\bar{L} - L_Y - L_K\right) = \rho\mu_4 - \dot{\mu}_4 \tag{4.65}$$

The transversality conditions are:

$$\lim_{t\to\infty} e^{-\rho t} \mu_1 K_t = 0 \tag{4.66}$$

$$\lim_{t\to\infty} e^{-\rho t} \mu_2 S_t = 0 \tag{4.67}$$

$$\lim_{t\to\infty} e^{-\rho t}\mu_3 Q_{K_t} = 0 \tag{4.68}$$

$$\lim_{t\to\infty} e^{-\rho t}\mu_4 Q_{E_t} = 0 \tag{4.69}$$

Equations (4.58), (4.59), (4.60), (4.61), (4.62), (4.63), (4.64), (4.65), (4.66), (4.67), (4.68) and (4.69) lead to the proposition below:

Proposition 4.2 In a centrally planned economy described by Eq. (4.56), the equilibrium growth rates of the variables in steady state are:

$$g_Y^{ss} = g_C^{ss} = g_K^{ss} = \frac{a_3\lambda\eta_E\overline{L} - (a_1 + a_3)\rho}{1 - a_2} \tag{4.70}$$

$$g_E^{ss} = -\rho \tag{4.71}$$

$$g_{Q_K}^{ss} = \lambda\eta_K L_K^{ss} \tag{4.72}$$

$$g_{Q_E}^{ss} = \lambda\eta_E\left(\overline{L} - L_Y^{ss} - L_K^{ss}\right) \tag{4.73}$$

where L_Y^{ss} and L_K^{ss} satisfy:

$$L_Y^{ss} = \frac{a_1\rho}{a_3\lambda\eta_E} \tag{4.74}$$

$$L_K^{ss} = \frac{a_3(a_1 + a_3)\lambda\eta_E^2\overline{L} - a_1(a_1 + a_3)\rho}{a_3(a_1 + a_3)\lambda\eta_E^2 + a_3^2\lambda\eta_K\eta_E} \tag{4.75}$$

As the steady-state growth rate of energy intensity $g_{In}^{ss} = g_E^{ss} - g_Y^{ss}$, Eqs. (4.70) and (4.71) in Proposition 4.2 inform that the steady-state growth rate of energy intensity will decline only if a positive rate of ESTC is present in the economy. That is to say, in centrally planned equilibrium, only energy R&D investment enables energy intensity improvements and sustainable growth.

4.3.3 Comparative Statics Analysis

Following the competitive equilibrium approach, I use the parameters of ρ(the subjective discount rate of representative households), $\overline{L}$ (aggregate level of labor supply) and η_E (incremental output of the ESTC R&D sector) to rewrite Eqs. (4.63), (4.64), (4.65), (4.66), (4.67) and (4.68):

$$g_Y^{ss} = \frac{a_3}{1-a_2}\lambda\eta_E\overline{L} - \frac{a_1+a_3}{1-a_2}\rho \tag{4.76}$$

$$g_E^{ss} = -\rho \tag{4.77}$$

$$L_Y^{ss} = \frac{a_1}{a_3\lambda\eta_E}\rho \tag{4.78}$$

$$L_K^{ss} = \frac{a_3(a_1+a_3)\lambda\eta_E^2\overline{L}}{a_3(a_1+a_3)\lambda\eta_E^2 + a_3^2\lambda\eta_K\eta_E} - \frac{a_1(a_1+a_2)}{a_3(a_1+a_3)\lambda\mu_E + a_3^2\lambda\eta_K\eta_E}\rho \tag{4.79}$$

$$g_{Q_K}^{ss} = \frac{a_3(a_1+a_3)\lambda^2\eta_K\eta_E^2\overline{L}}{a_3(a_1+a_3)\lambda\eta_E^2 + a_3^2\lambda\eta_K\eta_E} - \frac{a_1(a_1+a_3)\lambda\eta_E}{a_3(a_1+a_3)\lambda\eta_E + a_3^2\lambda\eta_K\eta_E}\rho \tag{4.80}$$

$$g_{Q_E}^{ss} = \frac{a_3^2(a_1+a_3)\lambda^2\eta_K\eta_E^2\overline{L}}{a_3(a_1+a_3)\lambda\eta_E^2 + a_3^2\lambda\eta_K\eta_E} - \frac{a_1(a_1+a_3)\lambda\eta_E}{a_3(a_1+a_3)\lambda\eta_E^2 + a_3^2\lambda\eta_K\eta_E}\rho - \frac{a_1}{a_3}\rho \tag{4.81}$$

Solve for their partial derivatives to obtain comparative statics analysis results, as shown in Table 4.5.

As shown in Table 4.5, in a centrally planned economy described by Eq. (4.56), the following inferences are valid in steady state:

Inference P-1 The steady-state economic growth rate g_Y^{ss} is a decreasing function of the subjective discount rate ρ, and an increasing function of the aggregate level of labor supply $\overline{L}$ and the incremental output of the ESTC R&D sector in steady state η_E.

Inference P-2 The steady-state growth rate of energy consumption g_E^{ss} is only a decreasing function of ρ.

Inference P-3 The labor employed by the final goods producing sector in steady state L_Y^{ss} is an increasing function of ρ, and a decreasing function of η_E.

Table 4.5 Comparative statics analysis of optimal economic equilibrium in centrally planned equilibrium

	$* = \rho$	$* = \overline{L}$	$* = \eta_E$
$\frac{\partial g_Y^{ss}}{\partial *}$	<0	>0	>0
$\frac{\partial g_E^{ss}}{\partial *}$	<0	=0	=0
$\frac{\partial L_Y^{ss}}{\partial *}$	>0	=0	<0
$\frac{\partial L_K^{ss}}{\partial *}$	<0	>0	<0
$\frac{\partial L_E^{ss}}{\partial *}$	<0	>0	>0
$\frac{\partial g_{QK}^{ss}}{\partial *}$	<0	>0	>0
$\frac{\partial g_{QE}^{ss}}{\partial *}$	<0	>0	>0

Inference P-4 The steady-state non-energy R&D investment in the intermediate goods producing sectors L_K^{ss} is a decreasing function of ρ and η_E, and an increasing function of $\overline{L}$.

Inference P-5 The steady-state energy R&D investment in the ESTC R&D sector L_E^{ss} is a decreasing function of ρ, and an increasing function of $\overline{L}$ and η_E.

Inference P-6 The steady-state rate of quality improvement of intermediate goods $g_{Q_K}^{ss}$ is a decreasing function of ρ, and an increasing function of $\overline{L}$ and η_E.

Inference P-7 The steady-state rate of ESTC $g_{Q_E}^{ss}$ is a decreasing function of ρ, and an increasing function of $\overline{L}$ and η_E.

4.3.4 Numerical Simulation Analysis

To make the results in Table 4.5 more intuitive and to compare them with the simulation results in competitive equilibrium, the same parameters as in numerical simulation in competitive equilibrium analysis are used to simulate the analysis in centrally planned equilibrium. Simulation results are shown in Table 4.6, the last column of which is the equilibrium rate of change in energy intensity, namely $g_{In}^{ss} = g_E^{ss} - g_Y^{ss}$. Figure 4.2 shows the equilibrium rate of ESTC and the equilibrium rate of change in energy intensity at different parameter values.

Table 4.6 Numerical simulation of optimal economic equilibrium in centrally planned equilibrium

a_1	a_2	a_3	ρ	$\overline{L}$	λ	η_K	η_E	L_Y^{ss}	L_K^{ss}	L_E^{ss}	g_Y^{ss}	$g_{Q_K}^{ss}$	$g_{Q_E}^{ss}$	g_{In}^{ss}
0.4	0.4	0.2	0.02	20	1	0.2	0.2	0.2	14.25	5.55	1.31	2.85	1.11	−1.33
0.4	0.4	0.2	0.04	20	1	0.2	0.2	0.4	13.5	6.1	1.29	2.7	1.22	−1.33
0.4	0.4	0.2	0.01	20	1	0.2	0.2	0.1	14.63	5.28	1.32	2.93	1.06	−1.33
0.4	0.4	0.2	0.02	30	1	0.2	0.2	0.2	21.75	8.05	1.98	4.35	1.61	−2
0.4	0.4	0.2	0.02	20	2	0.2	0.2	0.1	14.63	5.28	2.65	5.85	2.11	−2.67
0.4	0.4	0.2	0.02	20	0.5	0.2	0.2	0.4	13.5	6.1	0.65	1.35	0.61	−0.67
0.4	0.4	0.2	0.02	15	1	0.2	0.2	0.2	10.5	4.3	0.98	2.1	0.86	−1
0.4	0.4	0.2	0.02	20	1	0.4	0.2	0.2	11.4	8.4	1.31	4.56	1.68	−1.33
0.4	0.4	0.2	0.02	20	1	0.2	0.4	0.1	16.93	2.97	2.65	3.39	1.19	−2.67
0.4	0.4	0.2	0.02	20	1	0.4	0.4	0.1	14.81	5.09	2.65	5.93	2.04	−2.67
0.6	0.2	0.2	0.02	20	1	0.2	0.2	0.3	14.8	4.9	0.98	2.96	0.98	−1
0.2	0.6	0.2	0.02	20	1	0.2	0.2	0.1	13	6.9	1.98	2.6	1.38	−2
0.2	0.2	0.6	0.02	20	1	0.2	0.2	0.03	11.33	8.63	2.98	2.27	1.73	−3
0.4	0.3	0.3	0.02	20	1	0.2	0.2	0.13	13.53	6.33	1.69	2.71	1.27	−1.71

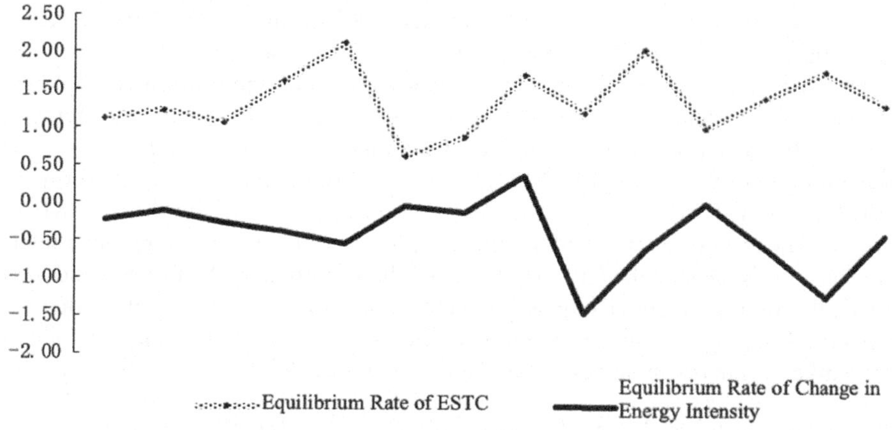

Fig. 4.2 Equilibrium rate of ESTC and equilibrium rate of change in energy intensity in centrally planned equilibrium

4.4 Analysis of Models When Elasticity of Substitution is not 1

4.4.1 Description of Economic Conditions

Sections 4.2 and 4.3 analyzed how R&D heterogeneity works in competitive equilibrium and in centrally planned equilibrium, under the assumption that $\sigma_{KLE} = \sigma_{K,E} = \sigma_{KE,L} = \sigma_{L,E} = \sigma_{LE,K} = \sigma_{K,L} = \sigma_{KL,E} \equiv 1$. This section will relax this assumption and focus on how the elasticity of substitution between capital and energy $\sigma_{K,E}$ will potentially affect the equilibrium rate of ESTC and the equilibrium growth path.

As mentioned in Sect. 4.1.3, the three-factor CES production function takes four forms. Given the specification of the production function and the importance of capital-energy substitution elasticity, I may assume, to simplify the analysis process, the elasticities of substitution other than the elasticity of substitution between capital and energy to be at certain values. For example:

$$Y = \left\{ (Q_E E)^{\frac{\sigma_{K,E}-1}{\sigma_{K,E}}} + \left[(Q_K K)^{\beta_1} L_Y^{1-\beta_1} \right]^{\frac{\sigma_{K,E}-1}{\sigma_{K,E}}} \right\}^{\frac{\sigma_{K,E}}{\sigma_{K,E}-1}} \quad (4.82)$$

$$Y = \left[(Q_E E)^{\frac{\sigma_{K,E}-1}{\sigma_{K,E}}} + (Q_K K)^{\frac{\sigma_{K,E}-1}{\sigma_{K,E}}} \right]^{\frac{\sigma_{K,E}}{\sigma_{K,E}-1}} \quad (4.83)$$

which respectively represent the forms of (KL)E, (KE)L, and (LE)K at certain values of the elasticity of substitution and other parameters.[2] Equation (4.82) is the form of (KL)E when $\sigma_{K,L} = 1$, which means that capital and labor are first employed to make intermediate goods through the Cobb–Douglas function, before energy is added to produce final goods through the CES function.[3] If $\beta_1 = 1$, then Eq. (4.82) becomes identical with Eq. (4.83). Equation (4.83) represents the specific forms of (KE)L and (LE)K when $\sigma_{KE,L} = 1$, $\sigma_{KE,L} = 0$ or $\sigma_{L,E} = 0$. In the form of (KE)L, Eq. (4.83) can indicate a situation where capital and energy jointly produce intermediate goods through the CES function, and then combine with labor to yield final goods through the Cobb–Douglas production function or the Leontief production function. If $\sigma_{KE,L} = 1$, then Y, E and K in Eq. (4.85) respectively represent "output per worker", "energy input per worker" and "capital stock per worker"; If $\sigma_{KE,L} = 0$, then according to the Leontief model, $\left[(Q_E E)^{\frac{\sigma_{K,E}-1}{\sigma_{K,E}}} + (Q_K K)^{\frac{\sigma_{K,E}-1}{\sigma_{K,E}}}\right]^{\frac{\sigma_{K,E}}{\sigma_{K,E}-1}} = L_Y$, when equilibrium is reached, and E and K in Eq. (4.83) denote respectively total energy input and total capital stock, meaning that in equilibrium gross output is equal to the amount of labor input in the final goods producing sector. In addition, if we directly relax the Leontief production function assumption in Hassler et al. (2012), i.e., $\sigma_{K,E} \neq 0$, then the production function in Hassler et al. (2012) may be rewritten as:

$$Y = \left[(Q_E E)^{\frac{\sigma_{K,E}-1}{\sigma_{K,E}}} + \left(Q_K K^a L_Y^{1-a}\right)^{\frac{\sigma_{K,E}-1}{\sigma_{K,E}}}\right]^{\frac{\sigma_{K,E}}{\sigma_{K,E}-1}} \tag{4.84}$$

which differs from Eq. (4.82) only in the combination of Q_K and K. In fact, if we assume $a = \beta_1 = 1$, then Eqs. (4.84) and (4.83) are identical with Eq. (4.82).

In sum, to avoid sacrificing generality, I omit discussions on capital quality and forms of the production function, and assume the production function of the final goods producing sector to be Eq. (4.83). In addition, given the similarities between the results of competitive equilibrium analysis and those of centrally planned equilibrium analysis, I take only the centrally planned equilibrium approach, and focus on the possible effect of changes in capital-energy substitution elasticity in Eq. (4.83) on the equilibrium rate of ESTC and the equilibrium growth rate.

The optimal growth problem faced by central planners is represented by the following system of equations:

[2] To highlight the importance of the elasticity of substitution between capital and energy, I disregard the form of (KEL) here. In fact, if both sides of the (KEL) function are divided by L_Y, then the form of (KEL) is similar to Eq. (4.83). In addition, as the present book focuses on ESTC, it does not discuss the specific forms of capital quality in depth, but represents them with capital-augmenting technological change.

[3] If $\sigma_{K,L} = 1$ in Eq. (4.82), then $\sigma_{K,E}$ is equal to $\sigma_{KL,E}$.

$$
\begin{aligned}
&\max_{C_t} U = \int_0^{\infty} e^{-\rho t} \ln C_t dt \\
&s.t.\ Y_t = \left[(Q_{E_t} E_t)^{\frac{\sigma_{K,E}-1}{\sigma_{K,E}}} + (Q_{K_t} K_t)^{\frac{\sigma_{K,E}-1}{\sigma_{K,E}}} \right]^{\frac{\sigma_{K,E}}{\sigma_{K,E}-1}} \\
&Y_t = C_t + \dot{K}_t \\
&\dot{S}_t = -E_t \\
&\dot{Q}_{Kt} = \lambda \eta_K L_{Kt} Q_{Kt} \\
&\dot{Q}_{Et} = \lambda \eta_E L_{Et} Q_{Et} \\
&L_{Kt} + L_{Et} = \bar{L_{KE}} \\
&\text{where } C_t,\ K_t,\ S_t,\ E_t,\ L_{K_t},\ L_{E_t},\ Q_{K_t},\ Q_{E_t} \geq 0,\ K_0,\ S_0,\ Q_{K_0},\ Q_{E_0}.
\end{aligned}
\tag{4.85}
$$

4.4.2 *Effect of Elasticity of Substitution on Equilibrium Rate of ESTC*

Based on Eq. (4.83), (4.86) is valid:

$$
Q_E = \left(\frac{MPE}{MPK} \right)^{\frac{\sigma_{K,E}}{\sigma_{K,E}-1}} \left(\frac{E}{K} \right)^{\frac{1}{\sigma_{K,E}-1}} Q_K \tag{4.86}
$$

where *MPE* and *MPK* respectively represent the marginal productivity of energy and that of capital. Equation (4.86) may be viewed as a variant of the core equation in Acemoglu's study on directed technical change. According to Eq. (4.83), the elasticity of substitution $\sigma_{K,E}$ has a positive effect on ESTC Q_E, while the effects of $\frac{MPE}{MPK}$ and of the factor endowment ratio $\frac{E}{K}$ on ESTC Q_E depend on the value of $\sigma_{K,E}$. If $\sigma_{K,E} > 1$, then Q_E is positively correlated with $\frac{MPE}{MPK}$ and $\frac{E}{K}$; if $\sigma_{K,E} < 1$, then Q_E is negatively correlated with $\frac{MPE}{MPK}$ and $\frac{E}{K}$.

Taking logarithms of both sides of Eq. (4.86) and solving for their derivatives with respect to time delivers:

$$
g_{Q_E} = \frac{\sigma_{K,E}}{\sigma_{K,E}-1}(g_{MPE} - g_{MPK}) + \frac{1}{\sigma_{K,E}-1}(g_E - g_K) g_{Q_K} \tag{4.87}
$$

From Eq. (4.85) is derived Eq. (4.88):

$$
g_E^{ss} = (1 - \sigma_{K,E})\left(g_Y^{ss} - g_{Q_E}^{ss}\right) - \sigma_{K,E}\rho \tag{4.88}
$$

If $\sigma_{K,E} = 1$, then Eq. (4.88) collapses to Eq. (4.71); if $\sigma_{K,E} > 1$, g_E^{ss} is positively correlated with $g_{Q_E}^{ss}$, and is negatively correlated with g_Y^{ss}; if $\sigma_{K,E} < 1$, g_E^{ss} is negatively correlated with $g_{Q_E}^{ss}$, and is positively correlated with g_Y^{ss}. If $\sigma_{K,E} > 1$, a greater rate of ESTC can effectively boost the substitution of energy for capital, accelerating energy consumption; if energy and capital are mutually complementary, the rate of ESTC will reduce both energy consumption and capital. In fact, as $g_Y^{ss} = g_K^{ss}$ in steady state, different values of $\sigma_{K,E}$ lead to different correlations between g_E^{ss} and g_Y^{ss}.

Assume Eq. (4.87) is in steady state, substitute Eq. (4.88) into (4.87), and eliminate g_E^{ss} to get:

$$g_{Q_E}^{ss} = \frac{\sigma_{K,E}}{\sigma_{K,E} - 1} \frac{g_{Q_K}^{ss}}{g_{Q_K}^{ss} - 1} (g_Y^{ss} - \rho) \tag{4.89}$$

As shown by the above-mentioned turnpike theorem Eq. (4.42), the rightmost term of Eq. (4.89) is a positive constant, so Eq. (4.89) can be re-expressed as:

$$g_{Q_E}^{ss} \sim \frac{\sigma_{K,E}}{\sigma_{K,E} - 1} \frac{g_{Q_K}^{ss}}{g_{Q_K}^{ss} - 1} \tag{4.90}$$

Solve for the partial derivatives of $\sigma_{K,E}$ and $g_{Q_K}^{ss}$ on both sides of Eq. (4.90):

$$\frac{\partial g_{Q_E}^{ss}}{\partial \sigma_{K,E}} = (\sigma_{K,E} - 1)^{-2} (g_{Q_K}^{ss} - 1)^{-2} \frac{\partial g_{Q_K}^{ss}}{\partial \sigma_{K,E}} \tag{4.91}$$

$$\frac{\partial g_{Q_E}^{ss}}{\partial g_{Q_K}^{ss}} = -(g_{Q_K}^{ss} - 1)^{-2} \frac{\sigma_{K,E}}{\sigma_{K,E} - 1} \tag{4.92}$$

Equations (4.91) and (4.92) warrant the validity of the following propositions:

Proposition 4.3 In a centrally planned economy represented by Eq. (4.85), the effect of $\sigma_{K,E}$ on $g_{Q_E}^{ss}$ in steady state is positively related with the effect of $\sigma_{K,E}$ on $g_{Q_K}^{ss}$.

Proposition 4.4 In a centrally planned economy represented by Eq. (4.85), if $0 < \sigma_{K,E} < 1$, then the effect of $g_{Q_K}^{ss}$ on $g_{Q_E}^{ss}$ is positive; if $\sigma_{K,E} > 1$, then the effect of $g_{Q_K}^{ss}$ on $g_{Q_E}^{ss}$ is negative.

As Proposition 4.3 shows, the effect of the elasticity of substitution $\sigma_{K,E}$ on the rate of ESTC $g_{Q_E}^{ss}$ is similar to that on the quality of intermediate goods $g_{Q_K}^{ss}$. That is to say, the effects of $\sigma_{K,E}$ on $g_{Q_E}^{ss}$ and $g_{Q_K}^{ss}$ are symmetrical, and will accelerate ESTC while improving the quality of intermediate goods. According to Proposition 4.4, if capital and energy are mutually complementary, $g_{Q_K}^{ss}$ and $g_{Q_E}^{ss}$ change in the same direction; if they are mutually substitutable, $g_{Q_K}^{ss}$ and $g_{Q_E}^{ss}$ change in opposite directions. Although Proposition 4.3 does not reveal whether $\sigma_{K,E}$ has a positive or a negative effect on $g_{Q_E}^{ss}$, Propositions 4.3 and 4.4 do show that whether factors of production are mutually substitutable or whether they are complementary to one another and how fast the quality of capital improves are key determinants of the effect of $\sigma_{K,E}$ on $g_{Q_E}^{ss}$. As Eq. (4.85) indicates, the elasticity of substitution between capital and energy $\sigma_{K,E}$ is actually equal to that between $Q_K K$ and $Q_E E$, so that the effect of $g_{Q_K}^{ss}$ on $g_{Q_E}^{ss}$ varies with different values of $\sigma_{K,E}$.

In addition, the production function Eq. (4.83) leads to:

$$g_Y = \theta_{EY}\left(g_{Q_E} + g_E\right) + \theta_{KY}\left(g_{Q_K} + g_K\right) \tag{4.93}$$

where θ_{EY} and θ_{KY} respectively denote the shares of energy and capital in output, which are defined in Eq. (3.6). Given that $\theta_{EY} + \theta_{KY} = 1$ and $g_Y^{ss} = g_K^{ss}$, I take the steady-state values on both sides of Eq. (4.93), substitute Eq. (4.90) into (4.93) to replace g_E^{ss}, and solve for the partial derivatives of $\sigma_{K,E}$ on both sides of the equation and substitute it into Eq. (4.91), and substitute $\frac{\partial g_{Q_K}^{ss}}{\partial \sigma_{K,E}}$ in Eq. (4.91) to obtain:

$$\frac{\partial g_{Q_E}^{ss}}{\partial \sigma_{K,E}} = \frac{\left(\partial \theta_{EY}^{ss}/\partial \sigma_{K,E}\right)\left(1/\theta_{EY}^{ss}\right) - 1/\sigma_{K,E} + \theta_{EY}^{ss}/\sigma_{K,E}}{\sigma_{K,E}\theta_{EY}^{ss} - \left(1-\theta_{EY}^{ss}\right)\left(\sigma_{K,E}-1\right)^2\left(g_{Q_K}^{ss}-1\right)^2} g_{Q_K}^{ss} \tag{4.94}$$

Based on Eq. (4.94), the following proposition is valid:

Proposition 4.5 In a centrally planned economy represented by Eq. (4.85), if $g_{Q_K}^{ss} > 0$, then $\frac{\partial g_{Q_E}^{ss}}{\partial \sigma_{K,E}} > 0$ when the following conditions are met:

$$\frac{\partial \theta_{EY}^{ss}}{\partial \sigma_{K,E}} > \frac{\theta_{EY}^{ss}\theta_{KY}^{ss}}{\sigma_{K,E}}, \text{ and } \frac{(\sigma_{K,E}-1)^2(g_{Q_K}^{ss}-1)^2}{\sigma_{K,E}+(\sigma_{K,E}-1)^2(g_{Q_K}^{ss}-1)^2} < \theta_{EY}^{ss} < 1;$$

$$\text{or } \frac{\partial \theta_{EY}^{ss}}{\partial \sigma_{K,E}} < \frac{\theta_{EY}^{ss}\theta_{KY}^{ss}}{\sigma_{K,E}}, \text{ and } 0 < \theta_{EY}^{ss} < \frac{(\sigma_{K,E}-1)^2(g_{Q_K}^{ss}-1)^2}{\sigma_{K,E}+(\sigma_{K,E}-1)^2(g_{Q_K}^{ss}-1)^2}.$$

If $g_{Q_K}^{ss} < 0$, then $\frac{\partial g_{Q_E}^{ss}}{\partial \sigma_{K,E}} > 0$ when the following conditions are met:

$$\frac{\partial \theta_{EY}^{ss}}{\partial \sigma_{K,E}} > \frac{\theta_{EY}^{ss}\theta_{KY}^{ss}}{\sigma_{K,E}}, \text{ and } 0 < \theta_{EY}^{ss} < \frac{\left(\sigma_{K,E}-1\right)^2\left(g_{Q_K}^{ss}-1\right)^2}{\sigma_{K,E}+\left(\sigma_{K,E}-1\right)^2\left(g_{Q_K}^{ss}-1\right)^2};$$

$$\text{or } \frac{\partial \theta_{EY}^{ss}}{\partial \sigma_{K,E}} < \frac{\theta_{EY}^{ss}\theta_{KY}^{ss}}{\sigma_{K,E}}, \text{ and } \frac{\left(\sigma_{K,E}-1\right)^2\left(g_{Q_K}^{ss}-1\right)^2}{\sigma_{K,E}+\left(\sigma_{K,E}-1\right)^2\left(g_{Q_K}^{ss}-1\right)^2} < \theta_{EY}^{ss} < 1.$$

4.4.3 Effect of Substitution Elasticity on Equilibrium Growth Rate

According to the de La Grandville hypothesis, a high value of substitution elasticity helps lift per capita income. Palivos and Karagiannis (2010) prove that the marginal productivity of factors of production will not decline to zero when the elasticity of substitution is greater than 1, so that sustainable growth can be generated endogenously. Xiaoling and Yujun (2013) validate the de La Grandville hypothesis with empirical data on China, thus confirming its validity in China. However, Grandville (2009), Palivos and Karagiannis (2010), Xiaoling and Yujun (2013) all miss the following questions. First, if the CES function includes energy, is the foregoing conclusion still valid? Second, how do changes in substitution elasticity affect steady-state economic growth?

Using the approach of de La Grandville et al., and substituting energy for labor in the production function used by Grandville (2009), Palivos and Karagiannis (2010), and Xiaoling and Yujun (2013), we can easily prove that, whether energy enters the CES function or not, a greater elasticity of substitution means higher income levels, and an elasticity of substitution greater than 1 can produce endogenous sustainable long-term growth. According to Uzawa (1961), under a neoclassical production function consisting of only capital and labor, a stable long-term growth rate requires that technological change in the production function be labor-augmenting or Harold-neutral, and the steady-state economic growth rate satisfy:

$$g_Y^{ss} \sim g_{Q_L}^{ss} \tag{4.95}$$

Therefore, if we prove that under Eq. (4.83) technological change in the production function is energy-augmenting when an economy grows steadily in the long run, then a variant of Eq. (4.95) could be obtained, which means that the steady-state growth rate is positively correlated with the steady-state rate of ESTC:

$$g_Y^{ss} \sim g_{Q_E}^{ss} \tag{4.96}$$

Uzawa (1961), Schlicht (2006), Jones and Scrimgeour (2008), Acemoglu (2008), and Irmen (2013) all discuss, and prove with different techniques, Uzawa's theorem under a neoclassical production function incorporating capital and labor, but they all neglect the energy input in their studies. According to Uzawa (1961), Schlicht (2006), Jones and Scrimgeour (2008), and Acemoglu (2008), if an economy in steady-state conforms to the balanced growth path, then we can assume that output, consumption, capital and energy all grow exponentially:

$$Y_t = Y_0 e^{g_Y t} \tag{4.97}$$

$$C_t = C_0 e^{g_C t} \tag{4.98}$$

$$K_t = K_0 e^{g_K t} \tag{4.99}$$

$$E_t = E_0 e^{g_E t} \tag{4.100}$$

where Y_0, C_0, K_0 and E_0 respectively denote output, consumption, capital and energy in the initial stage, and the initial value of each of the variables is greater than 0; g_Y, g_C, g_K and g_E, as in the above, mean the growth rates of respective variables. Re-express Eq. (4.11) in discrete form to make it clearer:

$$g_K K_t = Y_t - C_t \tag{4.101}$$

Substituting Eqs. (4.97), (4.98) and (4.99) into (4.101), and transposing the terms, we get:

$$g_K K_0 = Y_0 e^{(g_Y - g_K)t} - C_0 e^{(g_C - g_K)t} \tag{4.102}$$

Solve for the derivatives of both sides of Eq. (4.102) with respect to time:

$$(g_Y - g_K) Y_0 e^{(g_Y - g_C)t} - (g_C - g_K) C_0 = 0 \tag{4.103}$$

With regard to the relationships among g_Y, g_C and g_K in Eq. (4.103), if $g_Y = g_C$, Eq. (4.103) becomes $(g_Y - g_K)(Y_0 - C_0) = 0$, and $g_Y = g_K$ as K_0 is greater than 0; if $g_Y = g_K$, then $g_C = g_K$ in order to make Eq. (4.103) hold up. The validity of Eq. (4.103) means $g_Y = g_C = g_K$.

Since $G(E_t, K_t) \equiv Y_t = \left[\left(Q_{E_t} E_t\right)^{\frac{\sigma_{K,E}-1}{\sigma_{K,E}}} + \left(Q_{K_t} K_t\right)^{\frac{\sigma_{K,E}-1}{\sigma_{K,E}}} \right]^{\frac{\sigma_{K,E}}{\sigma_{K,E}-1}}$, $G(E_t, K_t)$ satisfies linear homogeneity:

$$Y_t = Y_0 e^{g_Y t} \equiv e^{g_Y t} G\left(E_t e^{-g_E t}, K_t e^{-g_K t}\right) = G\left(E_t e^{(g_Y - g_E)t}, K_t e^{(g_Y - g_K)t}\right) \tag{4.104}$$

where $E_t e^{-g_E t} = E_0$ and $K_t e^{-g_K t} = K_0$. Substitute the condition for the balanced growth path $g_Y = g_C = g_K$ into Eq. (4.104) to get the following propositions.

Proposition 4.6 In a centrally planned economy represented by Eq. (4.85), if output, consumption, capital and energy all grow exponentially according to Eqs. (4.97), (4.98), (4.99) and (4.100), then the technological change in the production function will be energy-augmenting when steady long-term economic growth is achieved.

Proposition 4.7 In a centrally planned economy represented by Eq. (4.85), if $g_{Q_K}^{ss} > 0$, then $\frac{\partial g_Y^{ss}}{\partial \sigma_{K,E}} > 0$ when the following conditions are met:

$$\frac{\partial \theta_{EY}^{ss}}{\partial \sigma_{K,E}} > \frac{\theta_{EY}^{ss} \theta_{KY}^{ss}}{\sigma_{K,E}}, \text{ and } \frac{(\sigma_{K,E} - 1)^2 (g_{Q_K}^{ss} - 1)^2}{\sigma_{K,E} + (\sigma_{K,E} - 1)^2 (g_{Q_K}^{ss} - 1)^2} < \theta_{EY}^{ss} < 1;$$

$$\text{or } \frac{\partial \theta_{\mathrm{EY}}^{\mathrm{ss}}}{\partial \sigma_{\mathrm{K,E}}} < \frac{\theta_{\mathrm{EY}}^{\mathrm{ss}} \theta_{\mathrm{KY}}^{\mathrm{ss}}}{\sigma_{\mathrm{K,E}}}, \text{ and } 0 < \theta_{\mathrm{EY}}^{\mathrm{ss}} < \frac{(\sigma_{\mathrm{K,E}} - 1)^2 (g_{\mathrm{Q_K}}^{\mathrm{ss}} - 1)^2}{\sigma_{\mathrm{K,E}} + (\sigma_{\mathrm{K,E}} - 1)^2 (g_{\mathrm{Q_K}}^{\mathrm{ss}} - 1)^2}.$$

If $g_{Q_K}^{ss} < 0$, then $\frac{\partial g_{Q_E}^{ss}}{\partial \sigma_{K,E}} > 0$ when the following conditions are met:

$$\frac{\partial \theta_{EY}^{ss}}{\partial \sigma_{K,E}} > \frac{\theta_{EY}^{ss} \theta_{KY}^{ss}}{\sigma_{K,E}}, \text{ and } 0 < \theta_{\mathrm{EY}}^{\mathrm{ss}} < \frac{\left(\sigma_{\mathrm{K,E}} - 1\right)^2 \left(g_{\mathrm{Q_K}}^{\mathrm{ss}} - 1\right)^2}{\sigma_{\mathrm{K,E}} + \left(\sigma_{\mathrm{K,E}} - 1\right)^2 \left(g_{\mathrm{Q_K}}^{\mathrm{ss}} - 1\right)^2};$$

$$\text{or } \frac{\partial \theta_{\mathrm{EY}}^{\mathrm{ss}}}{\partial \sigma_{\mathrm{K,E}}} < \frac{\theta_{\mathrm{EY}}^{\mathrm{ss}} \theta_{\mathrm{KY}}^{\mathrm{ss}}}{\sigma_{\mathrm{K,E}}}, \text{ and } \frac{\left(\sigma_{\mathrm{K,E}} - 1\right)^2 \left(g_{\mathrm{Q_K}}^{\mathrm{ss}} - 1\right)^2}{\sigma_{\mathrm{K,E}} + \left(\sigma_{\mathrm{K,E}} - 1\right)^2 \left(g_{\mathrm{Q_K}}^{\mathrm{ss}} - 1\right)^2} < \theta_{\mathrm{EY}}^{\mathrm{ss}} < 1.$$

4.4.4 Numerical Simulation Analysis

Consistent with Sects. 4.2 and 4.3, to make the analysis results more intuitive, I use numerical simulation in this section to simulate the effect of changes in major parameters on the equilibrium rate of ESTC and on the growth rate. Substitute Eq. (4.88) into (4.89), and do some trimming, to get:

$$g_{Q_E}^{ss} = \left[\frac{1 - \theta_{EY}^{ss}}{\theta_{EY}^{ss}} g_{Q_K}^{ss\ 2} - 2\rho\sigma_{K,E} g_{Q_K}^{ss} \right] \frac{1}{(\sigma_{K,E} - 1)\left(g_{Q_K}^{ss} - 1\right) - \sigma_{K,E} g_{Q_K}^{ss}} \tag{4.105}$$

Substitute Eq. (4.90) into (4.95) to obtain:

$$g_Y^{ss} = g_{Q_E}^{ss} + \frac{1 - \theta_{EY}^{ss}}{\sigma_{K,E} \theta_{EY}^{ss}} g_{Q_K}^{ss} - \rho \tag{4.106}$$

According to the definition of energy intensity, we can subtract Eq. (4.106) from (4.88) to get:

$$g_{In}^{ss} = -g_{Q_E}^{ss} - \frac{1 - \theta_{EY}^{ss}}{\theta_{EY}^{ss}} g_{Q_K}^{ss} \tag{4.107}$$

Equations (4.105), (4.106) and (4.107) form the foundation of numerical simulation in this section. As it is hard to obtain all steady-state analytical solutions under the CES production function, this section will, by selecting appropriate values of the subjective discount rate ρ, the elasticity of factor substitution $\sigma_{K,E}$, the energy share θ_{EY}^{ss} and the pace of improvement of capital quality $g_{Q_K}^{ss}$, simulate the effects of changes in these variables on the rate of ESTC $g_{Q_E}^{ss}$, the growth rate g_Y^{ss}, and the rate of change in energy intensity g_{In}^{ss}. The results of numerical simulation are shown in

Table 4.7. While the values of reference parameters in the table (the first line in Table 4.7) are taken from Wen and Zhaoxi (2012), the values of other parameters are based on Propositions 4.3, 4.4, 4.5, 4.6 and 4.7, and Inferences CES-1–CES-4 to simplify the calculation process. According to Inferences CES-1–CES-4, the reference value of θ_{EY}^{ss} is 0.5; the reference value of $\sigma_{K,E}$ is 1 when $\frac{(\sigma_{K,E}-1)^2(g_{QK}^{ss}-1)^2}{\sigma_{K,E}+(\sigma_{K,E}-1)^2(g_{QK}^{ss}-1)^2}$ is minimum; the reference value of g_{QK}^{ss} is 1 when $\frac{(\sigma_{K,E}-1)^2(g_{QK}^{ss}-1)^2}{\sigma_{K,E}+(\sigma_{K,E}-1)^2(g_{QK}^{ss}-1)^2}$ is minimum.

The simulation results in Table 4.7 are consistent with the basic conclusions of Propositions 4.3, 4.4, 4.5, 4.6 and 4.7. As Table 4.7 reveals, the elasticity of substitution and the share of energy in output combine to determine the results of comparative static analysis, which are as follows. Given that capital and energy are mutually complementary, a larger share of energy will lower the rate of ESTC in equilibrium when the quality of capital remains constant. Given that capital and energy are mutually substitutable, a smaller share of energy will increase the rate of

Table 4.7 Numerical simulation of optimal economic equilibrium

θ_{EY}^{ss}	$\sigma_{K,E}$	ρ	g_{QK}^{ss}	g_{QE}^{ss}	g_{Y}^{ss}	g_{E}^{ss}	g_{In}^{ss}
0.5	1	0.02	1	0.92	1.90	−0.02	−1.92
0.25	1	0.02	1	2.84	5.82	−0.02	−5.84
0.75	1	0.02	1	0.28	0.59	−0.02	−0.61
0.5	2	0.02	1	0.42	0.90	−0.52	−1.42
0.5	0.5	0.02	1	1.92	3.90	0.98	−2.92
0.5	1	0.04	1	0.84	1.80	−0.04	−1.84
0.5	1	0.01	1	0.96	1.95	−0.01	−1.96
0.5	1	0.02	2	1.92	3.90	−0.02	−3.92
0.5	1	0.02	1	0.92	1.90	−0.02	−1.92
0.5	1	0.02	0.8	0.72	1.50	−0.02	−1.52
0.5	2	0.02	2	1.23	2.21	−1.02	−3.23
0.5	2	0.02	0.8	0.28	0.66	−0.42	−1.08
0.5	0.5	0.02	0.4	−1.44	−0.66	0.38	1.04
0.5	0.5	0.02	0.3	−0.39	0.19	0.28	0.09
0.5	0.4	0.02	0.4	−0.74	0.24	0.58	0.34
0.5	0.1	0.02	0.8	−6.34	1.64	7.18	5.54
0.5	0.2	0.02	0.7	−4.79	−1.31	2.78	4.09
0.5	0.5	0.04	0.6	3.12	4.28	0.56	−3.72
0.25	2	0.02	1	1.34	2.82	−1.52	−4.34
0.75	2	0.02	1	0.11	0.26	−0.19	−0.45
0.25	0.4	0.02	0.5	−7.18	−3.45	2.23	5.68
0.75	0.4	0.02	0.5	−0.73	−0.33	0.23	0.56

ESTC in equilibrium when the quality of capital remains constant. Given that capital and energy are mutually complementary, better capital quality will elevate the rate of ESTC in equilibrium when the share of energy is constant. Given that capital and energy mutually substitutable, better capital quality will reduce the rate of ESTC in equilibrium when the energy share is constant. When other conditions are fixed, ESTC slows with rises in the subjective discount rate. When the sum of substitution elasticity and capital quality is 1, the rate of ESTC $g_{Q_E}^{ss}$, the growth rate g_Y^{ss}, and the rate of change in energy intensity g_{In}^{ss} will become extreme, because the denominator of Eq. (4.105) is 0 when the sum of substitution elasticity and capital quality is 1. With other conditions held constant, the rate of ESTC turns negative when the sum of substitution elasticity and capital quality is under 1, meaning that increases in the share of energy will accelerate ESTC in equilibrium. With other conditions held constant, the rate of ESTC is positive when the sum of substitution elasticity and capital quality is greater than 1, meaning that increases in the share of energy will retard ESTC in equilibrium.[4] With other conditions fixed, improvement of capital quality will invariably speed up ESTC.

Figure 4.3 shows how the equilibrium rate of ESTC and the equilibrium rate of change in energy intensity shift in an economy under the CES function Eq. (4.87). As shown by Fig. 4.3, as by Figs. 4.1 and 4.2, the equilibrium rate of ESTC and the equilibrium rate of change in energy intensity change in opposite directions. Equation (4.107) reveals that when the economic conditions depicted by Eq. (4.85) are in steady state, a positive rate of ESTC is a requisite for any improvement in energy intensity.

[4] Because $\sigma_{K,E} > 0$, if $\sigma_{K,E} + g_{Q_K}^{ss} < 1$, then $0 < \sigma_{K,E} < 1$ and $0 < g_{Q_K}^{ss} < 1$, or $\sigma_{K,E} > 1$ and $g_{Q_K}^{ss} < 0$. Suppose $\sigma_{K,E} + g_{Q_K}^{ss} \equiv x < 1$, as shown in Eq. (4.103), substituting x will turn Eq. (4.97) into $g_{Q_E}^{ss} = \left(\frac{1-\theta_{EY}^{ss}}{\theta_{EY}^{ss}} + 2\rho\right){g_{Q_K}^{ss}}^2 - 2\rho x$. If $0 < \sigma_{K,E} < 1$ and $0 < g_{Q_K}^{ss} < 1$, that means capital quality and ESTC are mutually complementary, so the equilibrium rate of ESTC will not be great. In addition, as $\sigma_{K,E}$ is also the elasticity of substitution between capital and energy, substituting Eq. (4.104) into (4.85) will lead to $g_E^{ss} \sim \frac{1-\sigma_{K,E}}{\sigma_{K,E}} \frac{1-\theta_{EY}^{ss}}{\theta_{EY}^{ss}} g_{Q_K}^{ss}$. Therefore, if $0 < \sigma_{K,E} < 1$ and $0 < g_{Q_K}^{ss} < 1$, then $g_E^{ss} < 0$. Because $g_K^{ss} = g_Y^{ss} > 0$, $g_{Q_E}^{ss} < 0$. When $g_{Q_K}^{ss} \to 1$, $g_{Q_E}^{ss}$ reaches its maximum value $\frac{1-\theta_{EY}^{ss}}{\theta_{EY}^{ss}} + 2\rho$. If $\sigma_{K,E} > 1$ and $g_{Q_K}^{ss} < 0$, that means capital quality and ESTC are mutually substitutable, so the equilibrium rate of ESTC and capital quality change in opposite directions. In addition, as $\sigma_{K,E}$ is also the elasticity of substitution between capital and energy, if $\sigma_{K,E} > 1$ and $g_{Q_K}^{ss} < 0$, then $g_E^{ss} < 0$. As the capital stock in steady state will be larger than energy, $g_{Q_E}^{ss} < 0$. When $g_{Q_K}^{ss} \to -\sigma_{K,E}$, $g_{Q_E}^{ss}$ reaches its maximum value $\left(\frac{1-\theta_{EY}^{ss}}{\theta_{EY}^{ss}} + 2\rho\right){\sigma_{K,E}}^2$.

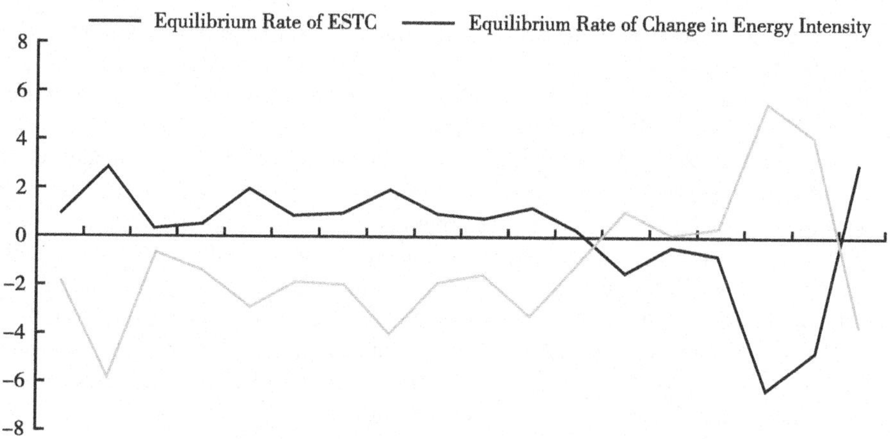

Fig. 4.3 The equilibrium rate of ESTC and the equilibrium rate of change in energy intensity

4.5 Comparison of Modeling Results and Discussion

In this chapter, I assume that the knowledge production function for the heterogeneous R&D sectors is linear. Although the assumption simplifies the solving process, it may lead to some problems. First, it misses the interaction between the heterogeneous R&D sectors. Second, the rate of technological change is constant and is irrelevant to the current state of technology. For this reason, I may relax the assumption based on different angles of study when I decide to expand the theoretical model in the future. To sum up, the model discussed in this chapter can be used as a benchmark for understanding endogenous ESTC in the context of R&D heterogeneity, whence I may attempt to do more as follows.

First, harmonization of the forms of a multi-factor nested CES function. As mentioned earlier, when two or more production factors are involved in a productive sector, the CES function takes at least four forms. Without the support of empirical studies, that may lead to problems in setting functional forms. The next step, therefore, should be to unify the forms of a multi-factor nested CES function into a "meta" function that satisfies the neoclassical production function assumption and contains different forms. From such a production function, we can obtain specific functional forms based on changes in parameters.

Second, establishment of production functions for the heterogeneous R&D sectors. As stated above, I use the assumption of the first-generation endogenous growth models to specify production functions for the heterogeneous R&D sectors. Under this assumption, the model built in this book, like the first-generation endogenous growth models, is faced with debate over two types of scale effect. Although neither semi-endogenous growth theory nor Schumpeterian growth theory can entirely solve the scale effect problem, it is still an objective worth working at. In addition, it is advisable to add spillovers between the heterogeneous R&D sectors into the model, as Pan Shiyuan and Shi Jinchuan (2002) have done. Alternatively, I may incorporate the competition-substitution relationship between heterogeneous R&D inputs into the model to discuss the effect of any changes in the substitution relationship on the final equilibrium results.

Third, introduction of renewable energy. In this book, I only consider non-renewable energy consisting of fossil fuels, thus leave room for inclusion of clean or renewable energy in the model. One treatment is to let E in Eq. (4.10) represent total energy input, assume that fossil fuels and clean energy are not fully substitutable, and respectively consider the impacts of fossil fuels- and clean energy-augmenting technological change and the allocation of R&D investment on the economic system. I may further consider the general equilibrium consequences of various R&D subsidy policies (such as subsidized fossil fuels- or subsidized clean energy-augmenting technological change), so as to assess relevant policies.

References

Acemoglu, D. (2008). *Introduction to Modern Economic Growth*. Princeton University Press.

Arrow, K. J., Chenery, H. B., Minhas, B. S., & Solow, R. M. (1961). Capital-labor substitution and economic efficiency. *The Review of Economics and Statistics, 43*(3), 225–250.

Dasgupta P. S., & Heal G. M. (1980). *Economic theory and exhaustible resources*. Cambridge: Cambridge University Press.

de la Grandville, O. (2009). *Economic Growth: A Unified Theory*. Cambridge University Press.

Dixit A. k., & Stiglitz J. E. (1977). Monopolistic competition and optimum product diversity. *The American Economic Review, 67*(3), 297-308.

Hassler, J., Krusell, P., & Olovsson, C. (2012). *Energy-saving technical change* (NBER Working Paper No. 18456).

Hotelling, H. (1931). The economics of exhaustible resources. *Journal of Political Economy, 39*(2), 137–175.

Irmen, A. (2013). *A generalized steady-state growth theorem* (CESifo Working Paper Series 4477), CESifo Group Munich.

Jones, C. I., & Scrimgeour, D. (2008). A new proof of Uzawa's steady-state growth theorem. *The Review of Economics and Statistics, 90*(1), 180–182.

Palivos, T., & Karagiannis, G. (2010). The elasticity of substitution as an engine of growth. *Macroeconomic Dynamics, 14*(5), 617–628.

Romer, P. M. (1990). Endogenous technological change. *Journal of Political Economy, 98*(5), S71–S102.

Sato, K. (1967). A two-level constant-elasticity-of-substitution production function. *Review of Economic Studies, 34*, 201–218.

Schlicht, E. (2006). A variant of Uzawa's theorem. *Economics Bulletin, 5*(6), 1–5.

Shiyuan, P., & Jinchuan, S. (2002). Endogenous growth theory: A literature review. *China Economic Quarterly, 03*, 753–786.

Solow, R. M. (1974). Intergenerational equity and exhaustible resources. *Review of Economic Studies, 41*(5), 29-45.

Stiglitz, J. E. (1974a). Growth with exhaustible natural resources: Efficient and optimal growth paths. *Review of Economic Studies, 41*(5), 123-137

Stiglitz, J. E. (1974b). Growth with exhaustible natural resources: The competitive economy. *Review of Economic Studies, 41*(5), 139-152.

Uzawa, H. (1961). Production function with constant elasticities of substitution. *Review of Economics Studies, 9*(4), 291–299.
Wen, X., & Zhaoxi, T. (2012). Renewable energy, intermediate goods quality and sustainable development. *Journal of World Economy, 02*, 143–159.
Xiaoling, C., & Yujun, L. (2013). Capital-labor substitution elasticity and regional economic growth: A test of the de La Grandville hypothesis. *China Economic Quarterly, 01*, 93–118.

Chapter 5
Estimation of ESTC Rates in China's Industrial Sectors: 1980–2010

In this chapter, I will relax the production function assumption introduced in Chap. 4, adopt a more general three-factor nested CES function to represent the behavior of the final goods producing sector, and make estimates about all forms of the three-factor nested CES production function. According to Eq. (5.3.5) in Chap. 3, ESTC is calculable if the form of the CES function is known. Estimating ESTC based on the three-factor nested CES function may lead to more credible results for the following reasons. Theoretically, estimating ESTC based on the three-factor nested CES function is a natural extension of the analysis in Chap. 4. Empirically, as the basic production function assumption about the final goods producing sector described in Chap. 4 is relaxed, it is possible to choose an appropriate method of estimation. Since the final goods producing sector requires the inputs of labor, capital and energy, I first discuss the methods of estimation for the CES function, compile and apply a sectoral input–output quantity and price database for the 1980–2010 period, and estimate the rates of factor-augmenting technological change and factor substitution elasticities in 37 industrial sectors. I then estimate the rates of ESTC in China's industrial sectors from 1980 to 2010 using Eq. (5.3.7) in Chap. 3.

5.1 Estimation Methods and Choice of Models

5.1.1 Comparison of Different Estimation Methods

The three-factor nested CES production function includes all possible forms of a production function with three production factors. Given the nested forms of the three-factor CES function in Chinese industry as shown in (4.3) and (4.4), I must estimate both the elasticity of substitution and factor-augmenting technological change before estimating ESTC. The existing literature records three major methods of estimating the nested CES function: single equation estimation, normalized supply-side

X. He, *The Endogenous Energy-Saving Technological Change in China's Industrial Sector*, https://doi.org/10.1007/978-981-19-7485-4_5

system approach, and first-order condition estimation. Below, I will discuss these methods and compare their merits and demerits.

(1) Single Equation Estimation

Kmenta (1967) initiated the single equation method of directly estimating the production function. He estimates the linear form of the production function by taking logarithms of the CES production function and using Taylor's formula for expansion around its unitary elasticity of substitution. However, as this method assumes that technological change in the production function is Hicks-neutral, namely, different production factors have the same form of factor-augmenting technological change, it has difficulty in estimating simultaneously elasticities of substitution and non-Hicks-neutral forms of factor-augmenting technological change. Lu Zhendong et al. (2009) use this method to estimate the three-factor nested CES production function in China. They find, upon comparing different elasticities of substitution, that all elasticities of substitution are significantly under 1, and the elasticity of substitution between capital and energy is far below the elasticities of substitution between other factors.

(2) Normalized Supply-Side System Approach

The normalized supply-side system approach was first introduced by Klump et al. (2007). They estimate the production function and its first-order conditions by normalizing the CES function.[1] León-Ledesma et al. (2010) proves by the Monte Carlo method that the normalized supply-side system approach is more effective than single equation estimation. However, when more than two inputs are involved, it is hard to make estimates with the parameters of a multi-factor nested CES function under this approach, because the price and quantity changes of a composite are unobservable. Therefore, current research is focused on how to estimate the CES function including only capital and labor by this method (Dai Tianshi & Xu Xianxiang, 2010; Chen Xiaoling & Lian Yujun, 2013). In addition, the choice of normalization methods will make a difference as to the estimation results.

(3) First-Order Condition Estimation

The first-order condition estimation method was invented by Edwin van der Werf (2008). The core of this method is to substitute the share changes of input factors for the price and quantity changes of a composite in a regression model so as to circumvent the problem of unavailability of the price and quantity changes of the composite. Therefore, it can be used to estimate multi-factor nested CES production functions. Making no assumption about the forms of technological change, Edwin van der Werf's (2008) method is generally applicable to various technological change assumptions, and can estimate both Hicks-neutral and non-Hicks-neutral technological change. De Cian (2009) also used this method to estimate multi-factor nested

[1] Normalization means indexation of inputs such as labor, capital and energy to minimize the effect of the dimensions of sample data on estimation results. See Klump et al. (2007) and de La Grandville (2009) for the normalization of two-factor CES functions, in which Klump et al. (2007) choose the averages of inputs and output, while de La Grandville (2009) chooses the initial inputs and output for normalization.

CES functions in OECD countries[2]; and Wei Wei & He Xubo (2014) applied this method to estimating the forms of the three-factor CES function in Chinese industry.

5.1.2 *Specification of the Estimation Model*

In this chapter, I will estimate the elasticities of substitution and factor-augmenting technological change of the three-factor nested CES function with capital, labor and energy as inputs. In view of the advantages and limitations of the foregoing three estimation methods, I use Edwin van der Werf's (2008) approach to estimate the three-factor nested CES production function in Chinese industry, and extend Wei Wei & He Xubo's (2014) work to all sectors of Chinese industry. Taking the (KL)E function as an example, I illustrate below how Edwin van der Werf (2008) generally estimates the three-factor nested CES function.[3]

In minimizing costs, representative manufacturers are faced with the following problem of planning:

$$\begin{aligned} &\min_{E,Z}: P_E E + P_Z Z \\ &s.t.\, \bar{Y} = [\alpha(Q_E E)^{\frac{\sigma_{KL,E}-1}{\sigma_{KL,E}}} + (1-\alpha)(Z)^{\frac{\sigma_{KL,E}-1}{\sigma_{KL,E}}}]^{\frac{\sigma_{KL,E}}{\sigma_{KL,E}-1}} \\ &\min_{K,L}: P_K K + P_L L \\ &s.t.\, \bar{Z} = [\beta(Q_K K)^{\frac{\sigma_{K,L}-1}{\sigma_{K,L}}} + (1-\beta)(Q_L L)^{\frac{\sigma_{K,L}-1}{\sigma_{K,L}}}]^{\frac{\sigma_{K,L}}{\sigma_{K,L}-1}} \end{aligned} \tag{5.1}$$

By (5.1), representative manufacturers first to solve for the optimal demand function for capital and labor when prices, technology and the quantity of the capital-labor composite $\bar{Z}$ are given, and then seek the optimal demand function for energy and the capital-labor composite when prices, technology and the quantity of the final good $\bar{Y}$ are given. Through this two-step optimization, the first-order conditions of representative manufacturers are obtained:

$$ln\left(\frac{E}{Y}\right) = \sigma_{KL,E} ln\alpha + (\sigma_{KL,E} - 1)\ln Q_E + \sigma_{KL,E} ln\left(\frac{P_Y}{P_E}\right) \tag{5.2}$$

$$ln\left(\frac{Z}{Y}\right) = \sigma_{KL,E} ln(1-\alpha) + \sigma_{KL,E} ln\left(\frac{P_Y}{P_Z}\right) \tag{5.3}$$

---

[2] Edwin van der Werf's (2008) estimated all forms of three-factor nested CES functions in OECD countries, while De Cian (2009) estimated only one specific form (KLE).

[3] Given that the present book is focused on ESTC, the empirical part omits discussions of the possible impact on capital quality, but concentrates on the possible techniques for estimating the rates of ESTC, as well as the possible effects of R&D heterogeneity on the rates of ESTC in China's industrial sectors.

$$ln\left(\frac{K}{Z}\right) = \sigma_{K,L} ln\beta + (\sigma_{K,L} - 1)\ln Q_K + \sigma_{K,L} ln\left(\frac{P_Z}{P_K}\right) \tag{5.4}$$

$$ln\left(\frac{L}{Z}\right) = \sigma_{K,L} ln(1-\beta) + (\sigma_{K,L} - 1)\ln Q_L + \sigma_{K,L} ln\left(\frac{P_Z}{P_L}\right) \tag{5.5}$$

where P_Y denotes the price of the final good. Performing first-order difference operations to (5.2), (5.3), (5.4), (5.5) leads to (5.6), (5.7), (5.8), (5.9):

$$e - y = (\sigma_{KL,E} - 1)q_E + \sigma_{KL,E}(p_Y - p_E) \tag{5.6}$$

$$z - y = \sigma_{KL,E}(p_Y - p_Z) \tag{5.7}$$

$$k - z = (\sigma_{K,L} - 1)q_K + \sigma_{K,L}(p_Z - p_K) \tag{5.8}$$

$$l - z = (\sigma_{K,L} - 1)q_L + \sigma_{K,L}(p_Z - p_L) \tag{5.9}$$

where k, l, e, z and y respectively represent changes in capital stock, labor input, energy consumption, the capital-labor composite and final output; q_K, q_L and q_E respectively stand for capital, labor and energy-augmenting technological change; and p_K, p_L, p_E and p_Z respectively denotes price changes in capital, labor, energy and the capital-labor composite.

As the quantity change z and the price change p_Z of the composite are not directly observable, it is hard to estimate the parameters if we construct an econometric model directly based on (5.6), (5.7), (5.8), (5.9). Edwin van der Werf's (2008) solution is to substitute the share changes of various factors (including the final good) for the quantity change z and the price change p_Z of the composite. Specifically:

First, adding $p_K - p_Y - (p_Z - p_Y)$ to both sides of (5.8), we obtain (5.10):

$$p_K + k - (p_Z + z) = (\sigma_{K,L} - 1)q_K + (\sigma_{K,L} - 1)[p_Z - p_Y - (p_K - p_Y)] \tag{5.10}$$

Secondly, adding $p_Z - p_Y$ to both sides of (5.7) results in (5.11):

$$p_Y - p_Z = \frac{p_Z + z - (p_Y + y)}{\sigma_{KL,E} - 1} \tag{5.11}$$

Finally, substituting (5.11) into the right-hand side of (5.10), we get:

$$p_K + k - (p_Z + z) = (\sigma_{K,L} - 1)q_K + (\sigma_{K,L} - 1)\left[\frac{p_Z + z - (p_Y + y)}{1 - \sigma_{KL,E}} - (p_K - p_Y)\right] \tag{5.12}$$

where the term on the left-hand side denotes the change in capital as a share of the composite, and $p_Z + z - (p_Y + y)$ on the right-hand side represents the change in the composite as a share of the final good. As the CES production function is a linear homogeneous equation, and Euler's law dictates that returns from the composite z are equal to the sum of returns from capital and labor, share changes can be calculated based on input prices and quantities.

Processing (5.9) in the same manner, we get the final system of equations:

$$e - y = (\sigma_{KL,E} - 1)q_E + \sigma_{KL,E}(p_Y - p_E) \tag{5.13}$$

$$\dot{\theta}_{KZ} = (\sigma_{K,L} - 1)q_K + \left(\frac{\sigma_{K,L} - 1}{1 - \sigma_{KL,E}}\right)\dot{\theta}_{ZY} + (1 - \sigma_{K,L})(p_K - p_Y) \tag{5.14}$$

$$\dot{\theta}_{LZ} = (\sigma_{K,L} - 1)q_L + \left(\frac{\sigma_{K,L} - 1}{1 - \sigma_{KL,E}}\right)\dot{\theta}_{ZY} + (1 - \sigma_{K,L})(p_L - p_Y) \tag{5.15}$$

where $\dot{\theta}_{mn} = p_m + m - (p_n - n)$ represents the change in the factor m as a share of n. By a similar method, we can obtain the final system of equations as follows in the forms of (KE)L and (LE)K production functions respectively:

$$l - y = (\sigma_{KE,L} - 1)q_L + \sigma_{KE,L}(p_Y - p_L) \tag{5.16}$$

$$\dot{\theta}_{KZ} = (\sigma_{K,E} - 1)q_K + \left(\frac{\sigma_{K,E} - 1}{1 - \sigma_{KE,L}}\right)\dot{\theta}_{ZY} + (1 - \sigma_{K,E})(p_K - p_Y) \tag{5.17}$$

$$\dot{\theta}_{EZ} = (\sigma_{K,E} - 1)q_E + \left(\frac{\sigma_{K,E} - 1}{1 - \sigma_{KE,L}}\right)\dot{\theta}_{ZY} + (1 - \sigma_{K,E})(p_E - p_Y) \tag{5.18}$$

$$k - y = (\sigma_{LE,K} - 1)q_K + \sigma_{LE,K}(p_Y - p_K) \tag{5.19}$$

$$\dot{\theta}_{EZ} = (\sigma_{L,E} - 1)q_E + \left(\frac{\sigma_{L,E} - 1}{1 - \sigma_{LE,K}}\right)\dot{\theta}_{ZY} + (1 - \sigma_{L,E})(p_E - p_Y) \tag{5.20}$$

$$\dot{\theta}_{LZ} = (\sigma_{L,E} - 1)q_L + \left(\frac{\sigma_{L,E} - 1}{1 - \sigma_{LE,K}}\right)\dot{\theta}_{ZY} + (1 - \sigma_{L,E})(p_L - p_Y) \tag{5.21}$$

Substituting new variables for those in (5.13), (5.14), (5.15) and adding random disturbance terms, we get the econometric models (5.22), (5.23), (5.24):

$$y_1 = \alpha_1 + \beta_1 x_1 + \varepsilon_1 \tag{5.22}$$

$$y_2 = \alpha_2 + \beta_{21} x_{21} + \beta_{22} x_{22} + \varepsilon_2 \tag{5.23}$$

$$y_2 = \alpha_3 + \beta_{31}x_{31} + \beta_{32}x_{32} + \varepsilon_3 \tag{5.24}$$

In order to keep the econometric model consistent with (5.13), (5.14), (5.15), we have to impose the following restrictions:

$$\beta_{22} = \beta_{32} \tag{5.25}$$

$$\beta_{21} = \beta_{31} = -\frac{\beta_{22}}{1-\beta_1} \tag{5.26}$$

Equations (5.22), (5.23), (5.24), (5.25) and (5.26) combine to form the econometric model under the (KL)E production function, and the econometric models under the (KE)L and (LE)K functions are similar to (5.22), (5.23), (5.24), (5.25) and (5.26). I will estimate the three models separately. Compared with the approaches of single equation estimation and normalized supply-side system estimation used in the relevant literature, the first-order condition approach provides great flexibility. Equations (5.22), (5.23), (5.24), (5.25) and (5.26) include both the CES production function in De Cian (2009) (the KLE function with only one layer of nesting) and Lu Zhendong et al.'s (2009) production function (which assumes only Hicks-neutral technological change). After estimating these models, I will test the CES assumptions and the Hicks-neutral technology assumption, namely: (1) if $\beta_{21} = \beta_{31} = 1$ and $\sigma_{KL,E} + \sigma_{K,L} = 0$, then the production function turns of KLE form; (2) if $\beta_1 = 1$, $\beta_{22} = \beta_{32} = 0$, and all elasticities of substitution are 1, then the production function becomes a Cobb–Douglas one; and (3) if $-\frac{\alpha_2}{\beta_{22}} = -\frac{\alpha_3}{\beta_{32}} = 0$, and $q_K = q_L = 0$, then the technological change in the production function becomes Hicks-neutral.

5.2 Choice of Variables, Data Sources and Pre-processing

5.2.1 Choice of Variables

I use Chen Shiyi's (2011) approach to measure sectoral output in terms of industrial value-added, and measure sectoral inputs in terms of capital stock, employment and total energy consumption.

5.2.2 Data Sources

The *Industrial Classification for National Economic Activities* ("GB/T4754"), which was first published and implemented in 1984, and was revised in 1994, 2002 and 2011, classifies industries and allocates industry codes in the Chinese economy. Since

2012, the latest 2011 edition (GB/T4754-2011) of industrial classification has been applicable to industrial statistics, though sectoral energy statistics are usually publicized with a two-year lag. Moreover, changes have been made to the classifications since 2012. For these reasons, industrial classification in this book is based on the 2002 edition ("GB/T4754-2002").

GB/T4754-2002 divides industry into three major categories: mining, manufacturing and the production and distribution of electricity, gas and tap water, under which there are 39 sectors and 191 groups. This book is concerned with the sectors (two-digit sectors). As the data about the sectors of "other ore mining and dressing" and "handicraft and other manufacturing" are minuscule, I merge these two sectors with the sector of "recycling and processing of waste resources and waste materials", on which data had not been available until 2003, into "others". See Table 5.1 for industrial classification.

To construct a sectoral input–output quantity and price database for 1980–2010, we have to grapple with two issues. First, different industrial classification standards employed by different issues of the statistical yearbook create data-matching problems. Second, inconsistent statistical standards preclude direct data comparison and analysis. The *China Statistical Yearbook 1987*, for instance, broadly categories industry into the metallurgical sector and the machinery sector and releases data on energy consumption in these sectors for 1980, 1985 and 1986. On the other hand, the *China Statistical Yearbook 1988* adopts a classification standard similar to the 1995 edition, except that (1) sectors 14–16 (See sector codes in Table 5.1) are collectively called "manufacture of foods, beverages and tobacco"; (2) coking and gas are crammed into one sector; and (3) sectors 35–41 are collectively named "manufacture of machinery, electrical and electronic equipment". The 1989–1994 issues of the *China Statistical Yearbooks* publicize data on industrial energy consumed from 1985 through 1992. Apart from the flaws found in the *China Statistical Yearbook 1988*, data for this period are lacking about sectors 8–10, 29, 20 and 46, and in the *China Statistical Yearbook 1988* sectors 17–19 are merged into "manufacture of textiles". After 1995, two adjustments are noteworthy: (1) the "mining" industry in the 1995–2002 issues of the *China Statistical Yearbook* includes "harvesting and transport of timber and bamboo", which is put under "agriculture" in later editions; and (2) the 2003 and later issues of the *China Statistical Yearbook* add a new sector—"recycling and processing of waste resources and waste materials"—under manufacturing.

The Explanation of Indicators part of the *Compilation of 50 Years' Statistical Data on China's Industry, Transport and Energy for 1949–1999* explains changes in the scope of Chinese industrial statistics. According to the Explanation, industrial surveys before 1997 were conducted on two groups of firms. One was all independent accounting industrial enterprises and non-independent accounting firms at or above the township level, on which data were collected through questionnaires. The other was industrial enterprises and household producers at or below the village level, including village-run manufacturers, urban cooperatives, rural cooperatives, urban household producers, and rural household producers (village-run manufacturers have not been included in industrial statistics until 1984), on which only global numbers of production were collected. Since 1998, the objects of industrial surveys have been

Table 5.1 The codes and names of the industrial two-digit sectors and 2010 energy consumption in Chinese industry

No	Sector code	Name of industrial two-digit sector	Ranking by energy consumption
1	6	Mining and washing of coal	7
2	7	Extraction of petroleum and natural gas	9
3	8	Mining and dressing of ferrous metal ores	20
4	9	Mining and dressing of non-ferrous metal ores	29
5	10	Mining and dressing of non-metallic ores	27
6	13	Processing of food from agricultural products	14
7	14	Manufacture of foodstuff	21
8	15	Manufacture of beverages	25
9	16	Processing of tobacco	35
10	17	Manufacture of textiles	8
11	18	Manufacture of clothing and fiber products	30
12	19	Manufacture of leather, fur, feather and related products	32
13	20	Processing of timber, manufacture of wood, bamboo, rattan, palm, and straw products	26
14	21	Manufacture of furniture	37
15	22	Manufacture of paper and paper products	10
16	23	Printing	33
17	24	Manufacture of articles for culture, education, and sports	36
18	25	Processing of petroleum and coking	5
19	26	Manufacture of chemical feedstock and chemical products	2
20	27	Manufacture of pharmaceuticals	24
21	28	Manufacture of chemical fibers	23
22	29	Manufacture of rubber	22
23	30	Manufacture of plastic products	17
24	31	Production of non-metallic minerals	3
25	32	Smelting and processing of ferrous metals	1

(continued)

Table 5.1 (continued)

No	Sector code	Name of industrial two-digit sector	Ranking by energy consumption
26	33	Smelting and processing of non-ferrous metals	6
27	34	Metalwork	12
28	35	Manufacture of general-purpose equipment	13
29	36	Manufacture of specialized equipment	18
30	37	Manufacture of transport equipment	11
31	39	Manufacture of electrical, machinery and equipment	16
32	40	Manufacture of computers and electronic communication equipment	15
33	41	Manufacture of instruments and meters, and cultural and office machinery	34
34	44	Production and distribution of electric power, heat power, and hot water	4
35	45	Generation and distribution of gas	31
36	46	Production and distribution of tap water	28
37	11, 42, and 43	Others	19

classified by scale, with all state-owned enterprises (SOEs) and non-state-owned industrial enterprises with annual sales in excess of 5 million yuan surveyed through questionnaires, and smaller-than-designated-size enterprises (non-SOEs with annual sales of less than 5 million yuan) surveyed through sampling. As informed by the explanation of indicators in the "Industry" chapter of the *China Statistical Yearbook*, industrial statistics 1998–2006 are about two groups: all SOEs and larger-than-designated-size non-SOEs, and smaller-than-designated-size firms, which from 2007 to 2010 were simply called larger-than-designated-size and smaller-than-designated-size industrial enterprises. Starting from 2011, however, only industrial enterprises making 20 million yuan or above of revenue from principal business per year qualify as "larger-than-designated-size" firms. Owing to a dearth of comparable data after 2010, my samples are limited to those for the 1980–2010 period.

5.2.3 Construction of a Database of Sectoral Input–Output Quantities and Prices: 1980–2010

In the current literature on energy use in China's industrial sectors, researchers approach the above-mentioned two problems—inconsistent industrial classification and statistical standards—generally by limiting samples to those following the same classification and statistical standards. For example, Li Xiaoping et al. (2008) select their samples from the 1998–2003 period; Tu Zhengge (2008) focuses on the 1998–2005 period; and Chen Yong & Li Xiaoping (2006) use samples from the 1985–2003 period. Chen Yong & Li Xiaoping's (2006) samples have the advantage of a long time span, but the problem of inconsistent statistical standards remains uncured. In the existing literature, only Chen Shiyi (2011) has done a fairly good job in making adjustments for different industrial classification and statistical standards, and has successfully constructed an input–output panel database ranging from 1980 to 2008 about China's 38 two-digit sectors. The input–output quantity and price database built in this book expands Chen Shiyi's (2011) work. This chapter uses directly the appendix to Chen Shiyi's (2011) paper, which includes data on value-added (with 1990 as the base year), capital stock (with 1990 as the base year) and employment in China's industrial sectors from 1980 to 2008, and builds a database of input–output quantities and prices from 1980 to 2008 by reference to Chen Shiyi's (2011) data processing principles and other related methods. Measures of inputs and output are total energy consumption, capital stock, number of employees and industrial value-added.[4]

[4] Processing total energy consumption data according to Chen Shiyi's (2011) principle may yield many abnormalities in the 1987–1992 statistics. For example, splitting data on energy use in the "manufacture of food" sector in 1992 delivers 3,988,400 TCEs, while the number for 1994 is 11,548,100 TCEs. The split number of energy use in the "tobacco processing" sector in 1992 is 14.2382 million TCEs, while the number for 1994 is 2.047 million TCEs. In the "clothing and other fiber products" sector, energy use figures balloon after 1987 (959,600 TCEs for 1986 but 3.636 million TCEs for 1987), with 6.5832 million TCEs for 1992 but 2.7623 million TCEs for 1994. In the "non-ferrous metal smelting and rolling" sector, 32.97 million TCEs in 1992 and 25.5512 million TCEs in 1994 are recorded. Starting from 1985, energy use figures about the sectors of "electrical, machinery and equipment manufacturing" and "electronic and communication equipment manufacturing" become unusually inflated (about the "electrical, machinery and equipment manufacturing" sector, the number of energy use is 7.161 million TCEs for 1980 and 8.12 million TCEs for 1984, but soars to 17.9556 million TCEs for 1985; the number about the "electronic and communication equipment manufacturing" sector is 1.8377 million TCEs for 1980 and 2.1437 million TCEs for 1984, while it is 5.038 million TCEs for 1985). The 1992 energy use numbers are 1.192.7 million TCEs in the "electrical, machinery and equipment manufacturing" sector and 8.047 million TCEs in the "electronic and communication equipment manufacturing" sector, while the 1994 numbers are 6.164 million TCEs and 3.7138 million TCEs respectively. With regard to those sectors still beset with statistical standards problems, I replace the data through linear interpolation, and smooth them by the moving average method. I also use linear interpolation to provide missing data, such as the data on energy use in all sectors in the 1980–1983 period and in 1993, in the sectors of "ferrous, non-ferrous and non-metallic mining and addressing", "wood processing", "printing", "manufacture of articles for culture, education, and sports", "rubber, plastic and non-metallic products", "other manufacturing" and "water production and distribution" from 1987 to 1993, in the sectors of "other ore mining and dressing", "food manufacturing", "tobacco processing", "electrical

(1) Sectoral Value-Added and Output Prices

(1) Sectoral Value-Added

I estimate the value-added of China's industrial sectors in 2009–2010 through three steps. (1) As the *China Statistical Yearbook* no longer publicizes data on the value-added of the industrial sectors after 2008, I calculate the nominal value-added of larger-than-designated-size enterprises in 2009–2010 using the value-added of 2007 and the growth rates from 2008 to 2010.[5] (2) I deflate the value-added to amounts in 1990 Chinese yuan based on the sectoral Producer Price Index (PPI), which is taken from the *2011 China Urban Life and Price Yearbook* (Table 2.3.3). (3) Based on the ratio of the gross industrial output from the *China Economic Census Yearbook 2008* to the gross output of larger-than-designated-size enterprises from the *China Statistical Yearbook 2008*, I adjust the data on sectoral value-added so that they are applicable to Chinese industry as a whole.

(2) Sectoral Output Prices

I use China's PPI, which comes from Table 2.3.3 in the *2011 China Urban Life and Price Yearbook*, as the sectoral output price index. As data are not available until 2006 on "other ore mining and dressing", and not available until 2002 on "processing of food from agricultural products", "printing and reproduction of recording media", "manufacture of general-purpose equipment", or "recycling of waste resources and waste materials", I take data from the *2011 China Urban Life and Price Yearbook* and use them as proxies for the missing data, substituting Table 2.3.1's "mining" for "other ore mining and dressing", and Table 2.3.2's "manufacture of foodstuff" for "processing of food from agricultural products", "manufacture of paper" for "printing and reproduction of recording media", "machinery" for "manufacture of general-purpose equipment", and "other industries" for "recycling of waste resources and waste materials". In view of the fact that the *China Urban Life and Price Yearbook* contains only data for 1985–2010, I use the sectoral PPI in the *China Statistical Yearbook 1993* as a proxy for data for 1980–1984. However, because the *China Statistical Yearbook 1993* applies a broad and fuzzy industrial classification standard (also applicable to the total energy consumption of the same period), I assume that different industries in the same sector have the same PPI.

machinery and equipment manufacturing" and "electronic and communication equipment manufacturing" from 1980 to 1993, in the sectors of "furniture manufacturing" and "metal products" in the 1980–1984 period and the 1987–1993 period, and in the sectors of "non-ferrous ore mining and dressing" and "non-metallic ore mining and dressing" sectors from 1980 to 1984. These outliers occur probably because of the inability of Chen Shiyi's (2011) splitting method to accurately process data on all the sectors, and because of his assumption, when he uses gross industrial output as the splitting weight, that there is a linear relationship between the total energy consumption and the gross output of a sector.

[5] See "The Sectoral Rates of Growth in Value Added" on the official website of the National Bureau of Statistics.

(2) Sectoral Capital Stock and Capital Goods Prices

(1) Sectoral Capital Stock

Generally, capital stock is estimated by the Perpetual Inventory Method (PIM). As Chen Shiyi (2011) has provided capital stock data for the 1980–2008 period, to estimate sectoral capital stock in 2009 and 2010, one needs to know the capital depreciation rate and new fixed investments for these two years. The common practice reported in the literature is to assume a fixed depreciation rate, such as the 9.6% depreciation rate assumed by Zhang Jun et al. (2004). Huang Yongfeng et al. (2002) assume that fixed capital comprises equipment and buildings, with the depreciation rate being 17% for equipment and 8% for buildings. Wang Yixuan & Wu You (2003) assume that fixed capital is made up of residential buildings, non-residential buildings, machinery and equipment, municipal structures, draught animals, rural houses and other fixed capital, with depreciation rates of 8%, 9%, 3.6–13.8%, 3.6%, 11% and 1.5% respectively. Wang Ling & Szirmai (2008) assume that fixed capital encompasses buildings and equipment, with buildings depreciating at 2.44%, and medical devices, electronic communication equipment and other equipment, under the category of equipment, depreciating at 8.11%, 9.45% and 8.99% respectively. Chen Yong & Li Xiaoping (2006), however, contend that those fixed depreciation rate assumptions are materially deficient, as different sectors may have different fixed asset structures, and it is difficult to determine the proportions of different types of fixed capital. Chen Shiyi (2011) estimates the depreciation rate using data on the accumulated depreciation, original value of fixed capital and net value of fixed capital of the industrial sectors taken from the past issues of the *China Industrial Economic Statistics Yearbook*:

$$\begin{aligned}\text{Accumulated depreciation}_t = {} & \text{Original value of fixed capital}_t \\ & - \text{Net value of fixed capital}_t\end{aligned} \tag{5.27}$$

$$\begin{aligned}\text{Depreciation of the year}_t = {} & \text{Accumulated depreciation}_t \\ & - \text{Accumulated depreciation}_{t-1}\end{aligned} \tag{5.28}$$

$$\begin{aligned}\text{Depreciation rate}_t = {} & \text{Depreciation of the year}_t / \\ & \text{Original value of fixed capital}_{t-1}\end{aligned} \tag{5.29}$$

The 2008–2011 issues of the *China Industrial Economy Statistical Yearbook* release data on the accumulated depreciation and the original value of fixed capital of larger-than-designated-size enterprises from 2007 to 2010, with the depreciation rate for the whole of industry calculated by using the coefficient for adjusting value-added.[6] In the literature, changes in the values of original and new fixed capital are usually used as proxies for investment outlays. Since changes in the original value of

[6] Calculated by Eqs. (5.27), (5.28), (5.29), the depreciation rate of the "manufacture of chemical fibers" sector in 2009 is −0.8%, which I suspect is due to wrong statistics. I therefore supplant it by the arithmetic mean of the depreciation rates of the sector in 2008 and 2010.

fixed capital of the "manufacture of chemical fibers" sector were negative in 2009, I use new fixed capital to represent the investment in the sector for that year. By reference to the fixed asset investment price index, I convert the fixed capital invested in 2009–2010 into an actual investment amount expressed in 1990 prices, which is adjusted to represent the investment in industry as a whole. Finally, I estimate the capital stock of each industrial sector from 2009 to 2010 by the Perpetual Inventory Method.

(2) Sectoral Capital Goods Prices

In the existing literature, capital goods prices are estimated mostly by three methods. The first is to use the fixed investment price index as the capital goods price index, as in Yang Zhongdong (2007, 2010). The second is based on the capital service price formula deduced by Christen & Jorgenson (1969), who assume that capital can be put into production or leased to generate interest, and that producers in possession of capital choose between production and leasing to minimize opportunity cost, as in Tao Xiaoma et al. (2009). The third method is to estimate the marginal rate of return on capital on the principle that marginal return on capital equals marginal cost, and to express the capital goods price as the marginal rate of return on capital, as in Lu Chengjun & Zhou Ruiming (2008), and Guo Juan et al. (2010).

Given the availability of data, I estimate the capital goods price by reference to the method adopted by Lu Chengjun & Zhou Ruiming (2008) and Guo Juan et al. (2010):

$$P(t) = \left[r(t) + \delta(t) - \frac{\dot{p}_k(t)}{p_k(t)} \right] (1 - f_\tau) p_k(t) \tag{5.30}$$

where $P(t)$ is the capital goods price in period t, $r(t)$ the nominal interest rate in period t, $\delta(t)$ the depreciation rate, $p_k(t)$ the capital market price in the year number t, $\frac{\dot{p}_k(t)}{p_k(t)}$ the expected rate of price change in the capital market, and f_τ the corporate income tax rate. Under the rational expectations assumption, $\frac{\dot{p}_k(t)}{p_k(t)}$ becomes the actual inflation rate $\pi(t)$, and as capital stock in China is measured in value rather than in kind, we may assume that $p_k(t) = 1$ (Guo Juan et al., 2010). Assuming $f_\tau = 0$, we turns the above equation into:

$$P(t) = r(t) + \delta(t) - \pi(t) \tag{5.31}$$

where $r(t)$ is represented by the one-year deposit interest rate for enterprises—in cases where several interest rates occur within one year, a time-weighted average applies, and the 1982 rate substitutes for the missing rates for 1980 and 1981; $\pi(t)$ is expressed as the GDP price deflator with 1990 as the base year; the depreciation rates of different sectors are derived from (5.29), but the depreciation rates for 1980–1985, 1993–2000 and 2004–2005 are incalculable due to data deficiency, necessitating the use of linear interpolation for those missing numbers.

(3) Sectoral Employment and Wages

(1) Sectoral Employment

Data on employment in China's industrial sectors in 2009 and 2010 are extracted from the 2010 and 2011 issues of the *China Statistical Yearbook*, adjusted by the ratio between the whole-of-industry employment data from the *China Economic Census Yearbook 2008* and those on larger-than-designated-size industrial firms for the corresponding years.

(2) Sectoral Wages

In the existing literature, Yang Zhongdong (2007, 2010) measures manufacturing labor cost by workers' average real wages, and creates a labor price index with 1978 as the base year. In calculating average wages, Yang Zhongdong (2007, 2010) uses data on average wages in manufacturing industries from the *China Statistical Yearbook*, with the data for 1979 and 1981–1983 derived by dividing total wages by total employees. Lu Chengjun & Zhou Ruiming (2008) employ the average wage index as the wage index for industry as a whole, and further calculate the real wages in various periods. Tao Xiaoma et al. (2009) arrive at the industrial labor price through dividing total wages in each sector by the number of employees in that sector. Guo Juan et al. (2010) treat the average wages of on-the-job workers as the industrial labor price, deflated by the consumer price index, to obtain the real labor price with 1978 as the base year, and use the index of average real wages of on-the-job workers as the real industrial labor price index. Chen Shiyi (2011) reminds us, however, that after SOEs laid off workers for the sake of efficiency in the late 1990s, measuring the labor input by the number of employees would substantially underestimate the labor input in production. To be in line with the employment data in Chen Shiyi (2011), I need first estimate the total real wages in various sectors.

As revealed by the past issues of the *China Labor Statistical Yearbook*, wage data published in 1999 and after are sectoral remuneration for all laborers paid by urban firms, those published in 1993–1999 are remuneration for permanent employees in industry, and those published in 1993 and before are total labor remuneration paid by entities of all forms of ownership (SOEs, urban collective-owned firms and other type enterprises). Assuming that the existing employment data on all forms of entities add up to the whole picture of industry, I obtain the total wage data for 1985–1992. In the 2002 and previous issues of the *China Labor Statistical Yearbook*, industry is broadly divided into mining, manufacturing, and the production and distribution of electricity, gas and tap water; Starting in 2003, breakdown data are available about the industrial two-digit sectors. Across the issues of the yearbook, inconsistent industrial classification and statistics standards remain an issue. To make workable the total wage data recorded in different issues, I need: (1) adjust the data after 1999, by a coefficient developed by Chen Shiyi (2011), so that they reflect all industrial sectors; (2) break down, following Chen Shiyi's approach, the headline wage data on mining, manufacturing and the production and distribution of electricity, gas, and tap water into data on the two-digit sectors; (3) supply the missing data for 1980–1984, 1986 and 1991 through linear interpolation; (4) deflate the nominal data by CPI (with 1990 as the base year) to obtain the total real wages for all sectors; and (5) divide the total real wages of each sector by the number of employees in that sector to get real wages per-worker as the sectoral labor price.

(4) Sectoral Total Energy Consumption and Energy Prices

(1) Sectoral Total Energy Consumption

In this book, I use total energy consumption as an energy consumption indicator. Energy consumption is a total quantitative indicator of the observed level, composition and growth rate of energy used by nationwide (or local) businesses and consumers in a certain period of time. Total energy consumption comprises three parts: end-use of energy, energy lost in processing and conversion, and other energy losses. Construction of a sectoral energy consumption database has already been described, so I will not repeat it here. Following is a description of how I build a sectoral energy price database.

(2) Sectoral Energy Prices

In the literature on energy prices, Yang Zhongdong (2007, 2010) uses the fuel and power purchase index ("FPPI") as the energy price index in manufacturing; but due to the lack of data, he substitutes the total retail price index of rural industrial products for the FPPI for 1978–1979, and the producer price index of coal for the FPPI for 1980–1988. Lu Chengjun & Zhou Ruiming (2008) represent the fuel and power price index as the energy price index, and estimate the total energy prices across the years after calculating the base-year price. Tao Xiaoma et al. (2009) obtain the energy price index by multiplying the fuel and power price indices in the *China Statistical Yearbook* and the earlier producer price indices in the coal, electricity and petroleum sectors, respectively, by the weights of the sectors' total output values; also they convert average prices into standard quantity prices using the 2003 energy price index in the table Average Prices in the Capital Goods Markets in 36 Large and Medium-sized Cities in the *China Price Yearbook 2004*, and calculate the 2003 average energy price weighted by the standard quantity of consumption of each energy source. Based on the proportion of coal in primary energy consumption in China, Guo Juan et al. (2010) obtain energy price data by reference to the average prices of coal sold by coal mines and the producer price indices of coal over the years, using the coefficient of conversion between raw coal and standard coal.

In Yang Zhongdong (2007, 2010) and Lu Chengjun & Zhou Ruiming (2008), it is assumed that all industrial sectors are charged the same price for energy consumption; and Guo Juan et al.'s (2010) method, if extended to all industrial sectors, makes it necessary to process coal consumption data in the same way as energy consumption data, especially, to split the existing data and supply the missing data. (For data processing purposes, I split the coal data on the same principle as the total energy consumption data, except that the 1980–1985 data include the leather sector, which is deemed as "manufacture of leather and feather products" in this book, and the textiles sector only comprises "textiles".) As shown by multi-year statistics, the proportion of coal in total energy consumption has been declining in most industrial sectors. For this reason, one can hardly accept Guo Juan et al.'s (2010) method of measuring the average energy price only by changes in coal prices.[7] Tao Xiaoma et al.'s (2009) approach seems more reasonable, though it does not specify the selected energy sources in citing the average energy prices in 36 large and medium-sized cities reported by the *China Price Yearbook 2004*, which may possibly lead to significant deviations in sectoral energy price calculations.[8] Nevertheless, Tao Xiaoma et al. (2009) and Guo Juan et al. (2010) have convinced me that calculating the sectoral average energy price requires two steps: firstly creating a sectoral energy consumption price index; secondly estimating the sectoral average energy consumption price for a certain year. Below is a description of how the sectoral average energy price is estimated following these two steps.

(a) Sectoral Energy Consumption Price Index

As electricity is generated mainly from thermal power in China,[9] I only take primary energy sources such as coal, oil and natural gas into account in developing the sectoral energy consumption price index. The index is constructed through the following steps: (1) I assume that the producer prices in the sectors of "mining and washing of coal" and "extraction of petroleum and natural gas" represent the price indices of coal and oil/natural gas. (2) As statistics on coal consumption from 1980 to 1985 are end use data, and the *China Energy Statistical Yearbook 1991* contains data on coal consumption from 1985 to 1988, I develop an adjustment coefficient by reference to the differences between the two data sets for 1985, and use it to convert data on end use of coal from 1980 to 1985 into total consumption data. (3) Coal consumption in each sector is converted into TCEs using a coefficient for converting different energy sources into TCEs provided by the *China Energy Statistical Yearbook*.[10] (4) As energy prices in different sectors in the 1980–1990 period are represented by prices in the three major industrial categories, I split coal consumption data following Chen

[7] Following Chen Shiyi's (2011) approach, I estimate the consumption of coal, oil and natural gas in industrial sectors, and find that coal consumption as a percentage of total energy consumption in most sectors displays a decreasing trend over the years. Take the mining and dressing of ferrous metal sector as an example. The proportion of coal consumption in this sector was 19.1% in 1990, fell to 12.8% in 2000 and 10.1% in 2010. In the "processing of petroleum and coking" sector, it was 12.5% in 1990, 15.8% in 2000 and 10% in 2010.

[8] In the *China Statistical Yearbook 2005*, Total Energy Consumption by Industry in 2003 is broken down into consumption of coal, coke, crude oil, gasoline, kerosene, diesel, fuel oil, natural gas, and electricity. In the *China Price Yearbook 2004*, Average Prices in the Capital Goods Markets in 36 Large and Medium-sized Cities in 2003 include clean coal for coking, mixed coal for industrial boilers, coke, No. 0 diesel oil, No. 10 diesel oil, No. 20 diesel oil, No. 93 unleaded gasoline, No. 90 unleaded gasoline, No.97 unleaded gasoline, heavy oil for fuel, general industrial electricity, industrial natural gas, and industrial coke oven gas.

[9] See published issues of the *China Electric Power Statistical Yearbook*.

[10] 1 ton of coal = 0.7143 TCE; 1 ton of oil = 1.4286 TCEs; 1,000 cubic meters of natural gas = 1.33 TCEs; 1 ton of coke = 0.9714 TCE; 1 ton of diesel = 1.4571 TCEs; 1 ton of gasoline = 1.4714 TCEs; 1 ton of fuel oil = 1.4256 TCEs.

Shiyi's (2011) approach, and allocate them to 37 industrial sectors. (5) I calculate the proportion of TCEs consumed by each sector in total energy consumption. (6) Given the preponderance of fossil fuels in primary energy consumption, I calculate energy consumption price indices for 34 sectors, excluding the sectors of "mining and washing of coal" and "extraction of petroleum and natural gas", by averaging the data mentioned in step (1) above weighted by the shares of coal, and oil and gas consumption in each sector. (7) I assume that only coal is consumed in the mining and washing of coal sector, so that energy prices are equivalent to coal prices in the sector. Likewise, I assume that only oil and natural gas are used in the sectors of the processing of petroleum, coking and gas generation and the extraction of petroleum and natural gas. (8) Outliers include data on the generation of electricity sector for 1980–1984, the sectors of mining and washing of coal, distribution of electricity and coking, and gas generation for 1985–1990, the mining and washing of coal sector for 1991–2010, the processing of tobacco for 1991–1992, the sectors of the processing of petroleum and coking and the manufacture of electronic communication equipment for 1991–1992, the distribution of electricity sector, and the generation and distribution of gas sector. To all those sectors mentioned above, smoothed moving average data apply.

(b) Sectoral Average Energy Prices in 2003

Based on Tao Xiaoma et al.'s (2009) approach, I assume that "coke" in the *China Price Yearbook 2004* is consistent with that in the *China Statistical Yearbook*, and classify "coking clean coal" and "blended coal for industrial boilers" in the price yearbook as "coal", and "fuel heavy oil" as "fuel oil". I add up the data on "No. 0", "No. 10" and "No. 20" diesel under the head of "diesel", and combine "No. 90", "No. 93" and "No. 97" unleaded gasoline into "gasoline" in the statistical yearbook. "Industrial natural gas" in the price yearbook is identical with "natural gas" in the statistical yearbook. Pursuant to the classification, the price of coal is expressed as the average of the "coking clean coal" and "blended coal for industrial boilers" prices; the price of fuel oil the average price of heavy oil for fuel; the price of diesel the average of the prices of three types of diesel; the price of gasoline the average of the prices of three types of gasoline; and the price of natural gas the average price of industrial natural gas.[11] Since the prices of coal, diesel and gasoline are averages of the averages, I may call the method "twice averaging".[12] Deflating by CPI the average prices to their 1990 values, calculating the proportions of coal, coke, natural gas, diesel, gasoline and fuel oil in total energy consumption[13] after converting them

[11] To make the figures comparable with those in the statistical yearbook, I multiply the price of natural gas by 10,000 to get "yuan/10,000 cubic meters of natural gas".

[12] I first calculate the average prices for 36 large and medium-sized cities, and then the average prices of different energy sources.

[13] The consumption of coal by the sectors of "mining and washing of coal", "production and distribution of electricity and heat" and "production and distribution of heat", converted into TCEs, still far exceeds the total energy consumption. To deal with the outliers about the three sectors, I express the proportion of coal consumption in each of these sectors as 1 minus "the proportions of coke, natural gas, diesel, gasoline and fuel oil in total energy consumption".

into TCEs with sectoral energy consumption data for 2003 in the *China Statistical Yearbook*, and solving for an average of the average prices of various energy sources weighted by their proportions, I obtain the average price for energy consumed by each sector in 2003.

With the price indices obtained through the method described above and the prices of energy used by different sectors in 2003, we can calculate the average energy prices in 37 industrial sectors from 1980 to 2010.

5.3 Analysis of Estimation Results

5.3.1 Estimated Elasticities of Substitution

I use a software application called Stata to estimate (5.22), (5.23), (5.24), (5.25), (5.26). I first substitute the estimated value of β_1 into (5.26), and then estimate (5.23) and (5.24) under the constraints of (5.25) and (5.26). I estimate the nested CES production function of each industrial sector in the forms of (KL)E, (KE)L and (LE)K respectively. Table 5.2 displays the estimated elasticities of substitution under different forms of CES production functions.

As revealed by Table 5.2, under the (KL)E functional form, estimated elasticities of substitution are significant at the level of 1%, except for the sectors of "plastic products", "production and distribution of electricity, steam and hot water" and "production and distribution of tap water". The estimated value of $\sigma_{KL,E}$ passes the significance test only in the sectors of "food", "clothing and fiber products", "fur, feather and related products", "printing", "articles for culture, education, and sports", "electronic, machinery and equipment", "computers and electronic communication equipment", and "production and distribution of gas". In those sectors where estimated elasticities of substitution are significant, $\sigma_{KL,E}$ is greater than $\sigma_{K,L}$, probably because the production processes in those sectors are less dependent on energy. Under the (KE)L functional form, estimated elasticities of substitution are significant at the level of 1% in most sectors, except in the sectors of "smelting and dressing of nonferrous metals", "metalwork", and "manufacture of general-purpose equipment", where estimated elasticities of substitution are significant at the level of 10%. The estimated value of $\sigma_{KE,L}$ is significant in most sectors, except in the sectors of "manufacture of beverages", "processing of tobacco", "manufacture of general-purpose equipment", "production and distribution of electricity, steam and hot water", and "production and distribution of tap water", which fail the significance test. In all of the sectors where estimated elasticities of substitution are significant, except the "manufacture of food" sector, $\sigma_{KE,L}$ is greater than $\sigma_{K,E}$. In addition, the distribution of $\sigma_{K,E}$ varies substantially across the sectors, with smaller estimated $\sigma_{K,E}$ in high energy-consuming sectors as a general rule. Under the (LE)K functional form, elasticities of substitution are significant only in the sectors of "mining and dressing of ferrous metal ores", "processing of petroleum and coking", "manufacture

Table 5.2 Estimated elasticities of substitution under different production functions

Sector No	(KL)E		(KE)L		(LE)K	
	$\sigma_{KL,E}$	$\sigma_{K,L}$	$\sigma_{KE,L}$	$\sigma_{K,E}$	$\sigma_{LE,K}$	$\sigma_{L,E}$
1	0.120	0.245***	0.254**	0.131***	0.034	0.999
2	2.530	0.508***	0.881***	0.093***	0.022	0.998
3	0.069	0.242***	0.346***	0.096***	0.005	0.944*
4	0.053	0.155***	0.180**	0.089***	0.005	0.979
5	0.014	0.164***	0.78***	0.360***	0.029	1.012
6	0.051	0.247***	0.782***	0.674***	0.018	0.996
7	0.523*	0.235***	1.088***	1.028***	0.004	0.994
8	0.028	0.153***	0.281	0.400***	0.006	0.991
9	0.109	0.364***	0.810	0.493***	0.042*	0.978
10	0.177	0.185***	0.931***	0.968***	0.002	1.002
11	0.313*	0.152***	0.359*	0.472***	0.026	0.971
12	0.527*	0.115***	0.531***	0.489***	0.023	0.973
13	0.381*	0.20 4***	0.386***	0.528***	0.012	0.985
14	0.034	0.193***	0.437***	0.681***	0.006	0.982
15	0.161	0.187***	0.417***	0.411***	0.008	0.997
16	0.420*	0.179***	0.397**	0.563***	0.028	0.99
17	0.420*	0.179***	0.396**	0.422***	0.003	0.951
18	0.115	0.244***	0.122***	0.179***	0.048	0.885***
19	0.081	0.097***	0.414**	0.069***	0.016	0.995
20	0.249	0.206***	0.549***	0.130***	0.012	0.975
21	0.049	0.200***	0.430**	0.033**	0.011	0.998
22	0.142	0.202***	0.583***	0.181***	0.012	0.99
23	0.193	0.235	0.457***	0.050***	0.019	0.995
24	0.031	0.175***	0.409***	0.146***	0.008	0.987
25	0.178	0.233***	0.324**	0.172***	0.024	0.998
26	0.157	0.203***	0.538***	0.023*	0.017	0.998
27	0.008	0.263***	0.287*	0.034*	0.011	0.987
28	0.064	0.184***	0.604	0.243*	0.003	0.993
29	0.088	0.149***	0.239*	0.360***	0.106	0.989
30	0.092	0.210***	0.314*	0.163***	0.037	0.993
31	0.216*	0.194***	0.651***	0.437***	0.012	0.944
32	0.268**	0.217***	0.420***	0.484***	0.055	0.906**
33	0.408	0.131***	0.384**	0.594***	0.024	0.991
34	0.273	0.935	0.289	0.258***	0.032	0.997
35	0.912*	0.547***	0.856***	0.806***	0.103	0.729***
36	0.042	0.998	0.322	0.668***	0.041	0.991

(continued)

Table 5.2 (continued)

Sector No	(KL)E		(KE)L		(LE)K	
	$\sigma_{KL,E}$	$\sigma_{K,L}$	$\sigma_{KE,L}$	$\sigma_{K,E}$	$\sigma_{LE,K}$	$\sigma_{L,E}$
37	0.293	0.182***	0.338**	0.239***	0.043	0.908*

Note *, ** and *** represent significance at the 10%, 5% and 1% levels respectively

of computers and electronic communication equipment", and "production and distribution of gas", all close to 1. The estimated value of $\sigma_{LE,K}$ is significant only in the "processing of tobacco" sector at the level of 10%, with factors of production being highly complementary. Looking closer at the significance levels of the different estimated elasticities of substitution in Table 5.2, one may find that the (KE)L functional form appears to be more likely to fit the conditions of Chinese industry.

5.3.2 Estimated Rates of ESTC in China's Industrial Sectors

When elasticities of substitution (Table 5.2) are estimated and the form of the CES production function is confirmed ((KE)L), one can estimate the rates of ESTC in various industrial sectors. To simplify the calculation process, I assume that $p = \frac{\sigma_{KE,L}-1}{\sigma_{KE,L}}$ and $p_1 = \frac{\sigma_{K,E}-1}{\sigma_{K,E}}$, and express the (KE)L functional form as:

$$Y = [\alpha(Q_L L)^p + (1-\alpha)(Z)^p]^{\frac{1}{p}} \tag{5.32}$$

$$Z = [\beta(Q_K K)^{p_1} + (1-\beta)(Q_E E)^{p_1}]^{\frac{1}{p_1}} \tag{5.33}$$

Solving for the first-order condition of E in (5.33) delivers:

$$\frac{\partial Z}{\partial E} = (1-\beta)Q_E^{p_1} E^{p_1-1} Z^{1-p_1} \tag{5.34}$$

To keep symbol consistency, θ_{EZ} denotes the share of energy in the capital-energy composite, and satisfies:

$$\theta_{EZ} = \frac{\partial Z}{\partial E}\frac{E}{Z} = (1-\beta)Q_E^{p_1} E^{p_1} Z^{-p_1} \tag{5.35}$$

Transposing the terms of (5.35), we obtain the expression of ESTC:

$$Q_E = \frac{Z}{E}\left(\frac{\theta_{EZ}}{1-\beta}\right)^{\frac{1}{p_1}} \tag{5.36}$$

Since the capital-energy composite Z in (5.36) is not directly observable, I need another observable variable in lieu of Z to calculate the rate of ESTC in each sector. The first-order condition of (5.32) is:

$$\frac{\partial Y}{\partial Z} = (1-\partial)Y^{1-p}Z^{p-1} \tag{5.37}$$

In the same manner, I calculate the proportion of the capital-energy composite in gross output:

$$\theta_{ZY} = \frac{\partial Y}{\partial Z}\frac{Z}{Y} = (1-\partial)Y^{-p}Z^{p} \tag{5.38}$$

Transposition of the terms of the equation leads to an expression of Z:

$$Z = Y\left(\frac{\theta_{ZY}}{1-\partial}\right)^{\frac{1}{p}} \tag{5.39}$$

Substituting (5.39) into (5.36) gives:

$$Q_E = \frac{Y}{E}\left(\frac{\theta_{ZY}}{1-\partial}\right)^{\frac{1}{p}}\left(\frac{\theta_{EZ}}{1-\beta}\right)^{\frac{1}{p_1}} \propto \frac{Y}{E}(\theta_{ZY})^{\frac{1}{p}}(\theta_{EZ})^{\frac{1}{p_1}} \tag{5.40}$$

where all the variables are observable, except the share parameters $1-\partial$ and $1-\beta$. The general practice in the existing literature is to calculate ESTC with the rightmost part of the above equation, without regard to the values of share parameters. This may be seriously problematic, however, for the factor shares θ_{ZY} and θ_{EZ} in (5.40) are unitless numbers, while output Y and energy E are measured in 100 million yuan and 100 million TCEs respectively. Calculation of ESTC Q_E by (5.40), therefore, leads to a result in the unit of "yuan per tonne", similar to the unit in which energy productivity is measured. As technological change should be unitless, (5.40) should be revised by the dimensionless method to solve the unit of ESTC.

There are two practical ways of making the ESTC equation dimensionless. One is to normalize (5.40) by dividing value-added Y and energy consumption E of each sector by the average values of the variables in that sector:

$$Q_E^* = \frac{Y/\overline{Y}}{E/\overline{E}}(\theta_{ZY})^{\frac{1}{p}}(\theta_{EZ})^{\frac{1}{p_1}} \tag{5.41}$$

The second way is to calculate ESTC using (5.40) directly:

$$q_E^* = y - e + \frac{1}{p}\dot{\theta}_{ZY} + \frac{1}{p_1}\dot{\theta}_{EZ} \tag{5.42}$$

where q_E^* is different from q_E. While q_E is a hypothetical constant rate of technological change for a simplified econometric model, q_E^* is an ESTC series changing with energy productivity and factor share. When the elasticities of substitution $\sigma_{KE,L}$ and $\sigma_{K,E}$ are estimated, (5.42) is capable of generating how the rate of ESTC changes in the sample period. Given the possible deficiencies of normalization (Temple, 2012) and the current productivity and rate of technological change used in empirical literature, I choose (5.42) to estimate the rates of ESTC in China's 37 industrial sectors from 1980 to 2010.

Table 5.3 shows the rates of ESTC and energy intensities in various industrial sectors in every ten years from 1981 to 2010. As demonstrated by the table, energy intensities are generally negatively correlated with ESTC rates, that is, higher rates of ESTC correspond with lower energy intensities. Specifically, in the sectors of the mining and dressing of ferrous metal ores, mining and dressing of non-ferrous metal ores, processing of food from agricultural products, manufacture of beverages, manufacture of paper and paper products, manufacture of articles for culture, education, and sports, production and distribution of electricity, steam and hot water, and production and distribution of gas, the rates of ESTC were negative in the 1981–2000 period, but they either turned positive or became much less negative in the 2001–2010 period, and the energy intensities improved significantly. On the other hand, in the sectors of the mining and washing of coal, extraction of petroleum and natural gas, processing of tobacco, manufacture of textiles, printing, processing of petroleum and coking, and production and distribution of tap water, the rates of ESTC turned from positive to negative in the 1981–2000 period, and the energy intensities in these sectors increased significantly. One may conclude that increased ESTC has made major contributions to the energy intensity improvements in Chinese industry.

The bottom lines of Table 5.3 record by sector grouping the simple arithmetic averages of the ESTC rates and energy intensities in various sectors. As shown, except for 1991–2000, the ESTC rates in high energy-consuming sectors and energy sectors not only exceeded those in low energy-consuming sectors and non-energy sectors, but also topped the all-sector average (from 1981 to 2000, ESTC in energy sectors was slightly slower than in non-energy sectors and all other sectors). In the whole sample period, ESTC in high energy-consuming sectors and energy sectors underwent a positive U-shaped change, which means that it started declining in the 1991–2000 period, but rising in the 2001–2010 period. Throughout the sample period, any industrial sector that experienced a higher ESTC rate for a certain period would see improved energy intensity compared with previous periods. That is to say, ESTC is a requisite for energy intensity improvement.

5.3.3 Comparison of Estimation Results

Comparison of the estimation results in Table 5.3 and the research of Lu Zhendong et al. (2009) reveals that Lu Zhendong et al. (2009) may have overlooked the heterogeneity of different sectors. For instance, Lu Zhendong et al. (2009) may

Table 5.3 Rates of ESTC and energy intensities in different industrial sectors: 1980–2010

Sector No	ESTC 1981–1990	Energy intensity 1980–1990	ESTC 1991–2000	Energy intensity 1991–2000	ESTC 2001–2010	Energy intensity 2001–2010
1	1.07	16.40	−2.70	17.60	3.96	10.60
2	1.15	7.82	−11.01	15.58	−0.52	18.81
3	−3.42	18.94	9.38	8.96	1.19	5.40
4	−5.20	9.63	9.45	5.56	0.88	2.82
5	4.93	3.35	4.21	3.09	2.73	3.08
6	−2.58	5.27	1.50	3.83	7.13	1.54
7	3.82	13.66	7.74	6.55	5.74	1.82
8	−2.64	5.24	11.22	2.83	4.36	0.96
9	3.74	0.57	−0.66	0.48	3.89	0.20
10	2.23	4.92	−1.56	3.75	4.54	2.13
11	3.66	1.24	5.86	1.13	5.26	0.58
12	3.08	3.24	9.11	1.80	7.16	0.62
13	9.53	5.87	15.63	3.86	5.04	1.07
14	0.71	2.64	9.30	1.57	7.47	0.48
15	−0.14	13.71	7.33	9.64	7.31	4.05
16	5.13	1.65	−5.16	1.66	4.75	0.73
17	−1.13	1.52	2.74	0.99	7.81	0.62
18	13.79	8.15	−14.91	28.08	3.49	41.03
19	1.56	21.07	5.20	20.37	17.17	8.61
20	8.74	12.07	12.34	3.61	5.61	0.76
21	7.99	13.54	1.82	6.64	−9.01	3.61
22	7.51	11.52	6.32	4.54	5.71	2.09
23	8.77	10.34	8.52	2.30	8.28	1.15
24	7.96	18.01	8.46	15.61	3.73	8.52
25	2.64	22.61	0.70	29.52	4.22	17.49
26	1.75	16.73	2.89	13.43	5.66	7.48
27	−0.54	4.44	6.87	2.57	3.17	1.68
28	0.07	5.55	9.49	2.57	6.15	0.91
29	−0.40	5.55	7.95	2.80	3.09	0.95
30	3.50	5.05	7.47	2.23	10.03	0.70
31	5.13	2.78	13.50	1.08	5.29	0.37
32	5.84	2.03	14.32	0.69	7.39	0 .18
33	1.10	5.70	13.24	1.40	7.98	0.34
34	−1.66	11.76	−2.42	15.60	7.17	10.67
35	−1.67	32.45	−32.06	97.92	19.18	14.68

(continued)

Table 5.3 (continued)

Sector No	ESTC 1981–1990	Energy intensity 1980–1990	ESTC 1991–2000	Energy intensity 1991–2000	ESTC 2001–2010	Energy intensity 2001–2010
36	3.57	14.35	−1.56	16.97	−0.41	17.35
37	−2.89	12.90	6.55	4.47	4.19	2.03
Industry as a Whole	2.61	9.52	3.97	9.76	5.32	5.30
High Energy-consuming Sectors	3.32	10.71	3.26	11.29	6.18	8.40
Low Energy-consuming Sectors	2.31	8.78	4.67	8.84	4.66	3.01
Energy Sectors	2.54	15.32	−12.62	34.96	6.66	19.16
Non-energy Sectors	2.63	6.62	6.57	5.83	5.12	3.14

have overestimated $\sigma_{K,E}$ and $\sigma_{KE,L}$ for high energy-consuming sectors, while underestimated $\sigma_{K,E}$ for low energy-consuming sectors. Apart from different estimation methods, those gaps in estimation results may be due to the following factors. Firstly, Lu Zhendong et al. (2009) employ macro data at the national level, which, relative to industrial production's dependence on input factors (especially energy), may underestimate the complementarity between input factors (especially between energy and capital). Secondly, the three-factor CES production function estimated in this chapter includes factor-augmenting technological change, while Lu Zhendong et al. (2009) only assume neutral technological change, without regard to the possible variant degrees of efficiency gains due to different production factors.

Limited by their methodology, Lu Zhendong et al. (2009) do not estimate the rate of factor-augmenting technological change in respect of each input factor. I compare the estimated values of a_K and a_L with the corresponding estimates in Dai Tianshi & Xu Xianxiang (2010) and Chen Xiaoling & Lian Yujun (2013), as shown in Table 5.4. As one can see in the table, the estimated rate of capital-augmenting technological change in this book is smaller than Dai Tianshi & Xu Xianxiang's (2010) and Chen Xiaoling & Lian Yujun's (2013) estimates, while my estimated rate of labor-augmenting technological change lies between their figures.

As indicated by the calculation results in Table 5.3, the rate of ESTC in each industrial sector is generally negatively correlated with the energy intensity in that sector from 1980 to 2010, which supports the credibility of the major propositions in Chap. 4.

Table 5.4 Comparison with Dai Tianshi & Xu Xianxiang (2010), and Chen Xiaoling & Lian Yujun (2013)

	Capital-augmenting technological change	Labor-augmenting technological change
This book	−0.060	0.068
Dai Tianshi & Xu Xianxiang (2010)	−0.011	0.093
Chen Xiaoling & Lian Yujun (2013)	0.019	0.033

Note The rates of capital- and labor-augmenting technological change in the "This book" line are estimated by obtaining their simple arithmetic averages in the sectors As Chen Xiaoling & Lian Yujun (2013) report in their book the estimated rates of capital- and labor-augmenting technological change by province/region, for comparison, I make simple arithmetic averages of their estimates

5.4 Robustness Testing

To ensure the robustness of the estimation results in Tables 5.2 and 5.3, I conduct hypothesis tests of the estimation results obtained under the (KE)L functional form, with the original hypotheses listed below and the test results shown in Table 5.5. In substance, the original hypothesis H3 is designed to test whether the rate of capital-augmenting technological change q_K and the rate of energy-augmenting technological change q_E can be zero simultaneously, while the rate of labor-augmenting technological change q_L significantly deviates from zero. Therefore, rejection of H3 does not in a strict sense guarantee that the rate of technological change in any sector is neutral, and may be because: (1) the rates of capital-, labor- and energy-augmenting technological change rates are all zero; or (2) the rate of labor-augmenting technological change is significantly different from zero, but the rates of capital- and energy-augmenting technological change are not zero simultaneously. Based on the H3 test results in Table 5.5, therefore, I have designed three other hypothesis tests to see whether technological change in any industrial sector is significantly neutral (see the original hypotheses H3-a, H3-b, and H3-c). I list the original hypotheses H1, H2, and H3 below, of which the specific forms are referred to in the "Specification of the Estimation Model" section of this chapter:

H1: The CES function is of KLE form;

H2: The CES function is a Cobb–Douglas one;

H3: The technological change in the CES function is neutral.

H3-a: The rates of capital- and labor-augmenting technological change are equal, namely, $q_K = q_L$;

H3-b: The rates of energy- and labor-augmenting technological change are equal, namely, $q_E = q_L$;

H3-c: The rates of capital- and energy-augmenting technological change are equal, namely, $q_K = q_E$.

Table 5.5 Hypothesis testing results under the (KE)L functional form

Sector No	(KE)L					
	KLE	Cobb–Douglas function	Neutral technological change	$q_K = q_L$	$q_E = q_L$	$q_K = q_E$
1	0.000	0.000	0.049	0.026	0.074	0.054
2	0.000	0.000	0.048	0.088	0.05	0.071
3	0.000	0.000	0.039	0.041	0.058	0.099
4	0.000	0.000	0.034	0.085	0.085	0.060
5	0.000	0.000	0.006	0.033	0.100	0.086
6	0.000	0.000	0.098	0.032	0.027	0.042
7	0.000	0.000	0.074	0.098	0.021	0.06
8	0.000	0.000	0.005	0.087	0.054	0.001
9	0.000	0.000	0.057	0.086	0.083	0.084
10	0.000	0.000	0.045	0.049	0.141	0.046
11	0.000	0.000	0.045	0.042	0.098	0.013
12	0.000	0.000	0.041	0.003	0.125	0.0 36
13	0.079	0.051	0.060	0.069	0.093	0.043
14	0.004	0.002	0.089	0.042	0.120	0.001
15	0.003	0.000	0.061	0.076	0.059	0.034
16	0.015	0.000	0.079	0.090	0.055	0.054
17	0.085	0.042	0.045	0.019	0.082	0.015
18	0.000	0.022	0.091	0.004	0.107	0.089
19	0.000	0.000	0.077	0.001	0.087	0.031
20	0.000	0.000	0.061	0.009	0.070	0.102
21	0.000	0.000	0.09	0.092	0.047	0.021
22	0.000	0.000	0.006	0.082	0.061	0.026
23	0.000	0.000	0.081	0.055	0.057	0.106
24	0.000	0.000	0.000	0.076	0.028	0.035
25	0.000	0.000	0.035	0.051	0.053	0.063
26	0.000	0.000	0.009	0.019	0.088	0.054
27	0.000	0.000	0.051	0.033	0.076	0.104
28	0.000	0.000	0.041	0.039	0.135	0.061
29	0.000	0.000	0.013	0.054	0.145	0.091
30	0.000	0.000	0.088	0.084	0.148	0.084
31	0.000	0.000	0.021	0.011	0.138	0.013
32	0.000	0.000	0.055	0.005	0.149	0.066
33	0.000	0.000	0.072	0.065	0.124	0.091
34	0.000	0.117	0.031	0.088	0.012	0.086

(continued)

Table 5.5 (continued)

Sector No	(KE)L					
	KLE	Cobb–Douglas function	Neutral technological change	$q_K = q_L$	$q_E = q_L$	$q_K = q_E$
35	0.051	0.097	0.096	0.109	0.093	0.108
36	0.000	0.001	0.002	0.092	0.116	0.071
37	0.000	0.000	0.019	0.086	0.062	0.004

Note The figures in the table are p-values of two-sided tests

As the tests show, except the "production and distribution of electricity, steam and hot water" sector, which cannot reject the original hypothesis on the Cobb–Douglas production function at the 10% significance level, all other sectors reject the original hypotheses on the KEL function, the Cobb–Douglas function and neutral technological change at the 10% significance level.

All sectors reject H3-a at the 10% significance level. Except the sectors of "textiles", "leather, feather and related products", "manufacture of furniture", "processing of petroleum and coking", "manufacture of general-purpose equipment", "manufacture of special equipment", "manufacture of transportation equipment", "manufacture of electrical, machinery and equipment", "manufacture of computers and electronic communication equipment", "manufacture of instruments and meters, cultural and office machinery" and "production and distribution of tap water", all other sectors reject H3-b at the 10% significance level, and reject the hypothesis $q_E = q_L$ at the 15% significance level. Except the sectors of "manufacture of pharmaceuticals", "manufacture of plastic products", "metalwork" and "production and distribution of gas", all other sectors reject the hypothesis $q_K = q_E$ at the 10% significance level, and the just-mentioned sectors join other sectors in rejecting H3-c at the 15% significance level.

Next, research should be focused on how to build samples and innovate estimation methods, including the following aspects.

First, construction of a sectoral input–output quantity and price database. Owing to the inconsistent statistical and industrial classification standards, a long-term sectoral input–output and price database is hard to build. In the existing literature, two approaches are often used to tackle this problem. One is to sacrifice sample size for data consistency; the other is to try to keep data consistency while using as many samples as possible. I follow the second approach, and rely heavily on the existing research. This has led to the problem that a large number of samples in the database referred to in this chapter are estimates, which are susceptible to errors. Should there exist any other innovative ways to process Chinese industrial statistics, we can compare them with Chen Shiyi's (2011) approach and learn from their strengths if any.

Second, the methods of estimating CES production functions. In this chapter, I have compared the merits of the three mainstream methods of estimating multi-factor

CES production functions and have chosen a relatively suitable one. I have to admit, however, all of them fail to consider the correlation between different industrial sectors. It is precisely because of this limitation that I regard ensuring a large sample size as a top priority in database construction.

Third, the constant elasticity of substitution assumption. Estimations in this chapter are based on the CES production function, which by assuming a constant elasticity of substitution immensely simplifies the analysis process. In the real world, the elasticity of substitution remains unchanged in the short run, but long-run economic data do not support this assertion (de La Grandville, 2009). Since the 1980s, Chinese industry has experienced several periods of leapfrog development; it is hardly convincing that the elasticity of substitution has been constant. Therefore, how to expand the existing econometric tools and estimate the varying elasticity of substitution effectively and accurately under the dual-constraints of limited samples and guaranteed degrees of freedom, should be the main hurdle we have to surmount in the next phase of study.[14]

References

Chengjun, L., & RuiDuanming, Z. (2008). Energy substitution in Chinese industry—Based on revisions of the Allen elasticity of substitution model. *The Journal of Quantitative & Technical Economics, 05*, 30–42.

De Cian, E. (2009). Factor-augmenting technical change: An empirical assessment. Fondazione Eni Enrico Mattei and School of Advanced Studies in Venice Working Paper, No.18.

de la Grandville. (2009). Economic growth: A unified theory. Cambridge: Cambridge University Press.

Juan, G., Chonghui, G., & Ling, Y. (2010). Energy rebound effect in Chinese industry. *The Journal of Quantitative & Technical Economics, 11*, 114–126.

Jun, Z., Guiying, W., & Jipeng, Z. (2004). Estimation of inter-provincial physical capital stock in China: 1952–2000. Economic Research Journal, 10, 35–44.

Klump, R., McAdam, P., & William, A. (2007). The normalized CES production function: Theory and empirics. European Central Bank Working Paper, No.1294.

Kmenta, J. (1967). On estimation of the CES production function. *International Economic Review, 8*(2), 180–189.

[14] The CES production function is used chiefly for the following reasons. Firstly, as revealed by the existing literature, the constant elasticity of substitution has been a common assumption in related research, because it can not only simplify the analysis process, but also fully describe economic activities. Secondly, from the perspective of basic production function theory, use of the VES production function may be problematic because no consensus has been reached on what factors will affect the elasticity of substitution, and whether the elasticity of substitution is endogenous or exogenous (i.e., whether it changes with time). Thirdly, from the perspective of econometric analysis, the CES production function is estimable. Moreover, whether the samples used in this book, i.e., China's industrial sectors, underwent so fundamental changes in the past 30 years from 1980 to 2010 that they affected the elasticity of substitution, merits further study. Though the constant elasticity of substitution assumption may not be convincing at present, little research has come out to verify or falsify it. In sum, controversial as it may be to some extent to describe the behavior of Chinese industry with the CES production function, it does not cause unacceptable deviations.

León-Ledesma, M. A., McAdam, P., & William, A. (2010). Identifying the elasticity of substitution with biased technical change. *American Economic Review, 100*(4), 1300–1357.

Ling, W., & Szirmai, A. (2008). The relationship between technology input and productivity growth in high-tech industry. *China Economic Quarterly, 03*, 913–932.

Shiyi, C. (2011). Estimation of statistical data on Chinese industrial sectors:1980–2008. *China Economic Quarterly, 03*, 735–776.

Temple, J. (2012). The calibration of CES production function. *Journal of Macroeconomics, 34*(2), 294–303.

Tianshi, D., & Xianxiang, X. (2010). The direction of China's technological progress. *Journal of World Economy, 11*, 54–70.

van der Werf, E. (2008). Production functions for climate policy modeling: An empirical analysis. *Energy Economics, 30*(6), 2964–2979.

Wei, W., & Xubo, H. (2014). Estimation of the energy CES production function in Chinese industry. *Journal of Beijing Institute of Technology (social Science Edition), 16*(1), 27–31.

Xiaoping, L., Xianxiang, L., & Zhongdi, Z. (2008). International trade, technological progress and productivity growth in Chinese Industry. China Economic Quarterly, 02, 549–564.

Xiaoling, C., & Yujun, L. (2013). Capital-labor substitution elasticity and regional economic growth: A test of the de La Grandville hypothesis. *China Economic Quarterly, 01*, 93–118.

Xiaoma, T., Jianwu, X., Xin, H., et al. (2009). Energy price distortion and factor substitution in Chinese industry. *The Journal of Quantitative & Technical Economics, 11*, 3–16.

Yixuan, W., & You, W. (2003). Preliminary calculation of the fixed capital stock of China's state-owned economy. Statistical Research, 05, 40–45.

Yong, C., & Xiaoping, L. (2006). Panel data construction and capital deepening evaluation in Chinese Industry: 1985–2003. *The Journal of Quantitative & Technical Economics, 10*, 57–68.

Yongfeng, H., Ruoen, R., & Xiaosheng, L. (2002). Estimation of the capital stock of China's manufacturing industry by perpetual inventory method. *China Economic Quarterly, 01*, 377–396.

Zhendong, L., Ju'E, G., & Youmin, X. (2009). Econometric estimation and selection of the CES production function for China's energy industry. China Population, Resources and Environment, 04, 156–160.

Zhengge, T. (2008). Coordination of the environment, resources and industrial growth. *Economic Research Journal, 02*, 93–105.

Zhongdong, Y. (2007). Energy substitution in China's manufacturing industry. Modern Economic Science, 03, 1–6+123.

Zhongdong, Y. (2010). Factors affecting the energy efficiency in China's manufacturing industry: Economic cycles and industrialization of the heavy chemical industry. *Statistical Research, 10*, 33–39.

Chapter 6
Empirical Analysis of R&D Heterogeneity-Based ESTC in China's Industrial Sectors

As the second part of my empirical analysis, this chapter builds on Chap. 5 to further examine the relationship between heterogeneous R&D investments and the rates of ESTC in China's industrial sectors, and test the basic assumptions in the framework constructed in this book. The major hypothesis put to test here is (4–16) suggested in Chap. 4, which states that investment in energy R&D has a positive effect on ESTC. Given that the existing statistics do not tell where R&D investment goes in the industrial sectors and of what nature it is, I need first to get clear on how energy R&D investment and non-energy R&D investment are measured, and then test empirically the relationship between R&D heterogeneity and the rates of ESTC through multiple econometric models.

6.1 Hypotheses

The major assumption in my framework of analysis is (4.16) in Chap. 4. As the existing data do not inform us of the destinations and natures of R&D investment in the industrial sectors, data on R&D investment in energy sectors (namely, the mining and washing of coal, extraction of petroleum and natural gas mining, processing of petroleum and coking, and production and distribution of electricity, heat and gas sectors) are taken as proxies for investment in energy R&D. This may underestimate the total R&D investment made by the whole of industry appropriated for energy efficiency gains for the following the reasons. Firstly, since energy sectors provide energy needed by other sectors, part of R&D investment in energy sectors is "reflected" in energy efficiency improvements in non-energy sectors. Secondly, non-energy sectors may improve energy efficiency through machine modification, staff training and learning by doing. Allowing for the limitations of the existing data, I recommend the following equation as an appropriate solution:

$$g_{Q_E}^{ss} \lambda \eta_E \bar{L} \tag{6.1}$$

X. He, *The Endogenous Energy-Saving Technological Change in China's Industrial Sector*, https://doi.org/10.1007/978-981-19-7485-4_6

which means that any increase in total R&D investment ($\bar{L}$) will affect the rate of ESTC, all else being equal. This leads to the following hypotheses.

Hypothesis 1: R&D investment is positively correlated with ESTC.

Although the directions of R&D investment in the industrial sectors are hardly detectable from statistical data, the results of the comparative static analysis in Chap. 4 show that new R&D investment in equilibrium is confined to energy R&D, for the gradual depletion of exhaustible resources broadens the return gap between heterogeneous R&D sectors. The more energy an industrial sector consumes, therefore, the greater the incentive becomes to improve energy efficiency in that sector.

If R&D investment in energy sectors and high energy-consuming sectors improves energy efficiency chiefly by increasing effective energy input, the degrees of improvement may be varied between the two groups of sectors for the same amounts of additional R&D investment. Based on how R&D investment is related to energy intensity in different sectors, we may obtain the following hypotheses.[1]

Hypothesis 2: R&D investment has a larger positive effect on ESTC in energy sectors than in non-energy sectors.

Hypothesis 3: R&D investment has a larger positive effect on ESTC in high energy-consuming sectors than in low energy-consuming sectors.

6.2 Choice of Variables

6.2.1 Explained Variable

The explained variable in this chapter, i.e., the rate of ESTC in each industrial sector, is derived from the estimation results in Chap. 5, denoted by *ESTC* below.

6.2.2 Explanatory Variables

The explanatory variables comprise the following two types of variable: R&D investment and sectoral characteristics.

---

[1] As energy intensity is mainly based on input–output analysis, it is merely an intuitive, not a comprehensive, indicator of energy efficiency. Its chief defects and the reasons behind them are illustrated in Chap. 3. In contrast, ESTC is a fairly comprehensive but abstract indicator and concept I propose in this book, in an attempt to overcome the shortcomings of other energy efficiency indicators. As I point out in Chaps. 4 and 5, ESTC is a requisite for the improvement of energy intensity, which should not be identified with ESTC itself. Moreover, as my empirical analysis shows, existing statistics only tell the distribution and evolution of energy intensity in various industrial sectors, but are not directly indicative of ESTC.

(1) R&D Investment

Internationally, R&D scale and R&D intensity are indicators of a country's scientific and technological strengths and core competencies. The R&D intensity of an industrial firm refers to the ratio of its R&D spending to its principal business revenue. R&D spending means a surveyed entity's actual expenses on internal R&D activities. By reference to the approach commonly applied in the relevant literature, I use the following indicators to measure R&D investment in an industrial sector: (1) R&D intensity, denoted by *RDin*; (2) R&D scale, i.e., total R&D expenditure, denoted by *RDscale*.

(2) Sectoral Characteristics

In this book, R&D heterogeneity includes energy R&D investment, i.e., investment in energy-related R&D activities, and non-energy R&D investment, i.e., R&D investment other than energy R&D investment. As the existing statistics offer no clue to energy-related R&D investment in each sector, I measure energy R&D investment by energy R&D spending in energy sectors (IEA, 2011; Chi, 2000; Chi et al., 2003a, b). Other non-energy sectors, especially high energy-consuming sectors, are also likely to invest in energy R&D, because energy accounts for a higher share of cost in high energy-consuming sectors than in low energy-consuming sectors (Hicks, 1932), and the fact that higher relative prices of energy can boost energy efficiency has long been recognized by economists (Ahmad, 1966). Drawing on the mainstream approach to measuring energy R&D investment and the theory of price-induced technological change, therefore, I use the nature of a sector and its level of energy consumption as dummy variables to define sectoral characteristics. To test Hypothesis 2 and Hypothesis 3, I have created dummy variables, named *HIGH* and *ENE,* to tell a high energy-consuming sector from a low energy-consuming sector, and an energy sector from a non-energy sector. I define high energy-consuming sectors and low energy-consuming sectors as per the rankings of industrial sectors by energy consumption in 2010 (Table 5.1). Energy sectors include the mining and washing of coal, extraction of petroleum and natural gas, processing of petroleum and coking, and production and distribution of electricity, heat and gas, as classified by the *China Statistical Yearbook*. All these sectors, except the production and distribution of gas, are high energy-consuming sectors. The settings of the dummy variables are as follows: HIGH = 1 means it is a high energy-consuming sector; HIGH = 0, a low energy-consuming sector; ENE = 1, an energy sector; ENE = 0, a non-energy sector.

6.2.3 Control Variables

In this chapter, I choose the following control variables, based on the empirical literature on the relationship between R&D investment and total factor productivity (TFP), as well as the theoretical and empirical literature on induced technological change. (1) Foreign direct investment (hereinafter denoted by "*FDI*"). As demonstrated by existing empirical studies, while FDI can crowd out R&D investment in a host country (Fan & Hu, 2007; Chengze et al., 2008; Yu & Guoliang, 2009), it helps increase the number of patent applications significantly (Hongling et al., 2006; Haiyang, 2008). Also, FDI can greatly reduce energy intensity in the host country (Xian & Yong, 2007). (2) Human capital (hereinafter denoted by "*HUM*"). Human capital is an important determinant of productivity growth (Engelbrecht, 1997; Barro, 2001; Caselli & Coleman, 2001). Liangke (2010), based on data on large and medium-sized industrial firms in China from 2000 to 2007, finds that human capital not only stimulates TFP in all sectors, but combined with R&D investment and R&D spillovers, has the function of an "assimilator". (3) Ownership structure (hereinafter denoted by "*SOE*"). The fuzzy boundaries of property rights and soft budget constraints drive underperformance and R&D inefficiency in SOEs (Genfu et al., 2006; Zhengge & Geng, 2006; Zhigang et al., 2006; Hongfeng & Zhaozhong, 2013), depressing the rates of ESTC of SOEs. (4) The spillover effect of energy quality (hereinafter denoted by "*RDspill*"). This variable refers to the phenomenon where R&D investment in energy sectors may crowd out that in non-energy sectors. Ran et al. (2010) and Xuan et al. (2012) rank it as a backward spillover effect or a backward linkage effect. (5) Energy prices (hereinafter denoted by "*PRICE*"). The theory of price-induced technological change suggests that rising energy prices are an important factor driving faster ESTC. (6) Capital stock (hereinafter denoted by "*CAP*"). As shown by the empirical results in Chap. 5, capital and energy are complementary to each other in China's industrial sectors.[2]

[2] According to Arrow (1962), learning by doing means that the unit cost of production decreases with the accumulation of production experience (e.g., output, installed capacity, and R&D investment). After Arrow (1962), the concept of learning by doing has been further expanded. Rosenberg (1982) introduces learning by using, and Cohen and Levinthal (1989) define learning by research. Kouvaritakis et al. (2000) combine learning by doing and learning by using to define the two-factor learning curve. As informed by the above-mentioned literature, the effect of learning by doing in a broad sense has a rich extent: learning by doing occurs whenever output or inputs accumulate in the production process. Given that *RDspill* and *CAP* respectively measure energy quality spillovers and capital stock in an industrial sector, if their value increases, one may say that that sector is experiencing a strong learning by doing effect. The control variables *RDspill* and *CAP* also control the effect of learning by doing on the rates of ESTC in China's industrial sectors to some extent.

6.3 Sources of Data, Data Preprocessing, and Descriptive Statistics of Variables

6.3.1 Sources of Data and Data Preprocessing

(1) Explained Variable

The rates of ESTC are derived from the estimation results in Chap. 5, spanning the 1981–2010 period.

(2) Explanatory Variables

Internal R&D spending data include those for the 1990–2010 period, extracted from the 1991–2011 issues of the *China Statistical Yearbook on Science and Technology*. For 2004 and 2009–2010, data on "other mineral mining and dressing" are missing; for 2002 and before, data on "timber and bamboo harvesting and transportation" are unnecessary. Data for 2004 and 2008–2009 are concerned with larger-than-designated-size industrial firms, and data for other years are about large and medium-sized industrial enterprises. Data on "principal business revenue" are only available for 2005–2010; for data that were not published in the statistical yearbook before 2005, I substitute value-added data. After converting by CPI the figures on internal R&D spending, principal business revenue and value-added into the money in 1990, we can directly calculate R&D intensity. For R&D investment in all sectors, the actual value of internal R&D spending needs to be adjusted by a coefficient consistent with value-added adjustment (see Chap. 5 for detailed adjustments). Finally, to ensure data stationarity, I take the natural logarithm of the actual amount of internal R&D spending.

(3) Control Variables

(1) Foreign Direct Investment

Borrowing Wenjun and Jinping's (2012) method, I measure FDI in a sector by the ratio of the original value of fixed investment made by solely-owned foreign enterprises, and Hong Kong, Macao and Taiwan firms to the total fixed investment in that sector, with the raw data coming from the *China Industry Economy Statistical Yearbook*. The fixed investment data, which only reflect the world of "SOEs and larger-than-designated-size industrial firms", need to be adjusted by the same coefficient for value added adjustment, as described in Chap. 5. In addition, the *China Industry Economy Statistical Yearbook* offers no data on fixed investment for 2004, and the 2000–2002 data exclude those on "handicraft and other manufacturing" and "recycling and processing of waste resources and waste materials"; for those missing data I substitute linearly fitted data. Since the FDI ratio in a sector is calculated using figures of the year concerned, no deflating is necessary to obtain FDI data for the 2000–2010 period.

(2) Human Capital

My calculation of human capital is inspired by Liangke (2010), who measures human capital by the share of personnel engaged in scientific and technological activities in the total labor force in an industrial sector. So far, the raw data on the scientific and technological personnel in each sector include those for 1990–2010, which come from the 1991–2011 issues of the *China Statistical Yearbook on Science and Technology*. Note that the data contain no information on "other mineral mining and dressing" for 2004 and 2009–2010, whereas the data for 2002 and before include those on the "timber and bamboo harvesting and transportation" sector, which do not exist for other years. The headcount in each sector is derived from the estimation results in Chap. 5. Given that the data for 2004 and 2008–2009 in the *China Statistical Yearbook on Science and Technology* concern larger-than-designated-size industrial enterprises, while those for other years deal with large and medium-sized industrial firms, I have, for the sake of accuracy, adjusted the number of scientific and technological personnel working in each sector by the same coefficient as with labor force adjustment, as detailed in Chap. 5.

(3) Ownership Structure

After the fashion of Hongfeng and Zhaozhong (2013), I define the ownership structure of a sector by the share of the total assets of state-owned and state-controlled industrial firms in the total assets of larger-than-designated-size industrial enterprises in that sector, with the raw data coming from the *China Statistical Yearbook*. No data on the "others" sector are available for 2000–2002, so I supply data for 2003–2005 as proxies for them. In the sectors of "tobacco products" and "processing of petroleum, coking and processing of nuclear fuel", the proportions of state-owned and state-controlled assets in those of larger-than-designated-size industrial firms for 2010 each exceed 1, apparently suggesting data errors in the statistical yearbook. For this reason, I use linearly fitted data in place of the raw data for 2010 about these two sectors. In addition, as the assets proportions are calculated using data for those years, I resort to no deflating operation to arrive at the true ownership structure of the industrial sectors for the 2000–2010 period.

(4) Spillover Effect of Energy Quality

I owe my treatment of the backward spillover or backward linkage effect to Ran et al. (2010) and Xuan et al. (2012). Let i and j respectively correspond to the row and column of the intermediate input matrix, which is denoted by $INT_{i,j}$, then $\omega_{i,j} = \frac{INT_{i,j}}{\sum_j INT_{i,j}}$ ($\omega_{i,j}$ means the element at row i and column j of the intermediate input matrix). If sector i is a non-energy sector, then $\omega_{i,j}$ represents the share of energy supplied by energy sector j in all energy that is consumed by non-energy sector i. Then R&D spillover from an energy sector to a non-energy sector can be expressed as:

$$RDspill_i = \sum_j RDspill_{i,j} = \omega_{i,j} RDscale_j \tag{6.2}$$

where $RDspill_{i,j}$ denotes R&D spillover from energy sector j to non-energy sector i; the total spillover from energy sectors to non-energy sectors *RDspill* can be obtained by putting together R&D spending in all energy sectors. $\omega_{i,j}$ in Eq. (6.2) is calculated based on the 2002 and 2007 editions of the *Input–Output Tables of China*. The spillovers in 2000–2005 are based on the 2002 edition of the *Input–Output Tables of China*, and those in 2006–2010 are based on the 2007 edition. The total spillover *RDspill* calculated by Eq. (6.2) needs no deflation because *RDscale* has already been deflated to the 1990 value, multiplied by the weighting factor $\omega_{i,j}$. Finally, I take the natural logarithm of the deflated total spillover *RDspill* to ensure data stationarity.

(5) Energy Prices and Capital Stock

As to how energy prices and capital stock are constructed, which is described in Chap. 5, I will say no more than that I take the natural logarithms of these variables for stationary data purposes in econometric analysis.

With only data on *FDI* and *HUM* for years after 2000 available, this chapter is focused on the relationship between R&D heterogeneity and ESTC in China's industrial sectors after 2000. The essential data in this chapter are the balanced panel data for 2000–2010, including 407 sample points.

6.3.2 Descriptive Statistics of Variables

Table 6.1 displays the descriptive statistics of all the variables. As revealed by the table, the maximum value of *ESTC* in Chinese industry is 0.720, its minimum value −0.503, its sample mean 0.052, and its sample standard deviation 0.117. The mean values of *ESTC* in high energy-consuming sectors and energy sectors are 0.061 and 0.069, respectively, higher than the mean value of 0.047 in low energy-consuming sectors and 0.051 in non-energy sectors. The standard deviations of *ESTC* in high energy-consuming sectors and energy sectors are also smaller than those in low energy-consuming sectors and non-energy sectors, respectively. The maximum value of RDin is 1.364, its minimum value 0.002, its sample mean 0.081, and its sample standard deviation 0.149. The maximum value of ln*RDscale* is 7.246, its minimum value −1.555, its sample mean 3.351, and its sample standard deviation 1.69. As the data show, R&D intensity in Chinese industry is more disparately distributed among different sectors than is R&D scale. In terms of sectoral characteristics, the mean values of *RDin* in high energy-consuming sectors and energy sectors are 0.119 and 0.157, respectively, higher than the 0.046 and 0.072 in low energy-consuming sectors and non-energy sectors, where differences in the distribution of *RDin* between sectors are also great. ln*RDscale* averages 4.503 in high energy-consuming sectors and 3.064 in energy sectors, and ln*RDscale* is higher in high energy-consuming sectors than the 2.259 in low energy-consuming sectors, but it is significantly lower in energy sectors than the 3.385 in non-energy sectors.

Table 6.2 shows the coefficients of correlation between *ESTC* and other variables in different sectors, with the coefficients denoted by the same symbol in different

Table 6.1 Descriptive statistics of major variables

		Sample size	Mean	Standard deviation	Minimum	Maximum	Median
ESTC	All sectors	407	0.052	0.117	−0.503	0.720	0.055
	High energy	198	0.061	0.107	−0.398	0.520	0.057
	Low energy	209	0.047	0.126	−0.503	0.519	0.043
	Energy sectors	44	0.069	0.111	−0.398	0.720	0.066
	Non-energy sectors	363	0.051	0.147	−0.503	0.519	0.055
RDin	All sectors	407	0.081	0.149	0.002	1.364	0.040
	High energy	198	0.119	0.189	0.012	1.364	0.065
	Low energy	209	0.046	0.084	0.002	0.714	0.022
	Energy sectors	44	0.157	0.238	0.002	1.364	0.082
	Non-energy sectors	363	0.072	0.132	0.004	1.255	0.038
ln*RDscale*	All sectors	407	3.351	1.690	−1.555	7.246	3.464
	High energy	198	4.503	1.128	2.159	7.246	4.398
	Low energy	209	2.259	1.382	−1.555	5.637	2.333
	Energy sectors	44	3.064	1.859	−1.555	5.628	3.587
	Non-energy sectors	363	3.385	1.668	−0.943	7.246	3.450
FDI	All sectors	407	0.285	0.194	0.000	0.796	0.287
	High energy	198	0.266	0.184	0.000	0.796	0.255
	Low energy	209	0.303	0.202	0.000	0.682	0.344
	Energy sectors	44	0.086	0.100	0.000	0.407	0.043
	Non-energy sectors	363	0.309	0.189	0.000	0.796	0.312
HUM	All sectors	407	0.017	0.015	0.001	0.079	0.011
	High energy	198	0.023	0.015	0.003	0.068	0.022
	Low energy	209	0.012	0.014	0.001	0.079	0.006
	Energy sectors	44	0.023	0.017	0.001	0.068	0.017
	Non-energy sector	363	0.017	0.015	0.001	0.079	0.009
SOE	All sectors	407	0.409	0.289	0.006	0.993	0.351
	High energy	198	0.457	0.282	0.037	0.989	0.434
	Low energy	209	0.363	0.289	0.006	0.993	0.293

(continued)

Table 6.1 (continued)

		Sample size	Mean	Standard deviation	Minimum	Maximum	Median
	Energy sectors	44	0.835	0.158	0.354	0.989	0.890
	Non-energy sectors	363	0.357	0.257	0.006	0.993	0.304
ln*SPILL*	All sectors	407	10.524	3.248	0.381	16.130	11.281
	High energy	198	12.392	2.088	5.288	16.130	12.84344
	Low energy	209	8.754	3.161	0.381	13.592	9.272582
	Energy sectors	44	12.533	1.737	8.455	16.130	12.86612
	Non-energy sectors	363	10.280	3.305	0.381	15.592	10.87022
ln*PRICE*	All sectors	407	5.901	0.894	3.491	8.807	5.818
	High energy	198	6.116	0.908	4.503	8.807	6.017
	Low energy	209	5.696	0.831	3.491	8.118	5.687
	Energy sectors	44	6.885	1.133	5.105	8.807	6.857
	Non-energy sectors	363	5.781	0.782	3.491	7.899	5.773
ln*CAP*	All sectors	407	6.897	1.134	3.784	9.941	6.886
	High energy	198	7.709	0.821	5.583	9.941	7.661
	Low energy	209	6.127	0.805	3.784	7.787	6.192
	Energy sectors	44	8.080	1.203	5.953	9.941	8.180
	Non-energy sectors	363	6.753	1.039	3.784	9.080	6.831

Note "High Energy" denotes high energy-consuming sectors, and "Low Energy" low energy-consuming sectors

sectors. *ESTC* is significantly positively correlated with all variables, except *SOE,* but its correlations with other variables vary from sector to sector. *ESTC*'s correlations with *RDin* and ln*RDscale* are much stronger in high energy-consuming sectors and energy sectors than in other sectors. In high energy-consuming sectors and energy sectors, *ESTC*'s correlations with ln*SPILL* and *HUM* are much significant than in other sectors, while its correlations with *FDI* and *SOE* are weaker than in low energy-consuming sectors and non-energy sectors.

Table 6.2 Coefficients of correlation between *ESTC* and other variables in different industrial sectors

	All sectors	High energy-consuming sectors	Low energy-consuming sectors	Energy sectors	Non-energy sectors
RDin	0.537***	0.658***	0.341***	0.667***	0.055***
ln*RDscale*	0.470***	0.588***	0.236***	0.597***	0.174***
HUM	0.361**	0.417***	0.254**	0.272***	0.146**
ln*SPILL*	0.427***	0.439***	0.103**	0.260***	0.070**
FDI	0.233**	0.174*	0.394***	0.266	0.123**
SOE	−0.278***	−0.516***	−0.131***	−0.656***	−0.090**
ln*PRICE*	0.067**	0.112*	0.261*	0.009*	0.111*
ln*CAP*	0.042**	0.043*	0.052*	0.308*	0.033*

Note *, ** and *** in the above table indicate that the variables concerned are significant at the 10%, 5% and 1% significance level respectively

6.4 Econometric Model Specification and Results of Empirical Analysis

6.4.1 Econometric Model Specification

To test the hypotheses I raise in this chapter, the specification of the econometric model consists of two parts: the regression of *ESTC* on *RDin*; and the regression of *ESTC* on ln*RDscale*. In the empirical analysis, I use the methods of mixed data regression, panel data regression, and dynamic panel data regression.

(1) Specification of Mixed Data Regression Models

The complete mixed data regression models are Eqs. (6.3) and (6.4):

$$\begin{aligned} ESTC_i &= a_0 + a_1 RDin_i + a_2 HIGH * RDin_i \\ &\quad + a_3 ENE * RDin_i + A^T X + \varepsilon_i \end{aligned} \tag{6.3}$$

$$\begin{aligned} ESTC_i &= b_0 + b_1 \ln RDscale_i + b_2 HIGH * \ln RDscale_i \\ &\quad + b_3 ENE * \ln RDscale_i + B^T X + \varpi_i \end{aligned} \tag{6.4}$$

where the subscript i denotes different values of observation; a_0 and b_0 the intercept terms of the regression models; a_i and b_i ($i \in \{1, 2, 3\}$) the regression coefficients of the explanatory variables; X the column vector consisting of different control variables; A and B the column vector consisting of the regression coefficients of different control variables, with the superscript T denoting the transpose; and ε and ϖ the random disturbance terms. The regression coefficients a_1 and b_1 can be regarded as the average effects of increases in R&D intensity (one unit) and R&D scale (one percent) on *ESTC*, respectively; and a_2 and b_2 can be regarded as the marginal effects of increases in R&D intensity (one unit) and R&D scale (one percent) on *ESTC* in high energy-consuming sectors. So the total effects in high energy-consuming sectors are $a_1 + a_2$ and $b_1 + b_2$ respectively. In the same vein, the total effects of increases in R&D intensity (one unit) and R&D scale (one percent) on *ESTC* in energy sectors are $a_1 + a_3$ and $b_1 + b_3$ respectively.

(2) Specification of Panel Data Regression Models

Before specifying panel data models, I choose between the fixed effects model and the random effects model by the Hausman test, and the test results favor the fixed effects model. Accordingly, I have constructed panel data fixed effects models Eqs. (6.5) and (6.6):

$$\begin{aligned} ESTC_{i,t} &= a_0 + a_1 RDin_{i,t} + a_2 HIGH * RDin_{i,t} \\ &\quad + a_3 ENE * RDin_{i,t} + A^T X + \mu_t + \mu_i + \varepsilon_{i,t} \end{aligned} \tag{6.5}$$

$$\begin{aligned} ESTC_{i,t} &= b_0 + b_1 \ln RDscale_{i,t} + b_2 HIGH * \ln RDscale_{i,t} \\ &\quad + b_3 ENE * \ln RDscale_{i,t} + B^T X + \mu_t + \mu_i + \varpi_{i,t} \end{aligned} \tag{6.6}$$

where the subscripts i and t denote sector and year respectively; μ_t and μ_i represent period and individual fixed effects; other variables and coefficients have the same meanings as in Eqs. (6.3) and (6.4).

(3) Specification of Dynamic Panel Data Regression Models

Given the possible inertia in technological change leading to current *ESTC* being correlated with its lagged values (Otto, 2006), I have attempted to model the dynamics of *ESTC* as follows:

$$\begin{aligned} ESTC_{i,t} &= aESTC_{i,t-1} + a_1 RDin_{i,t} + a_2 HIGH * RDin_{i,t} \\ &\quad + a_3 ENE * RDin_{i,t} + A^T X + \mu_t + \mu_i + \varepsilon_{i,t} \end{aligned} \tag{6.7}$$

$$\begin{aligned} ESTC_{i,t} &= bESTC_{i,t-1} + b_1 \ln RDscale_{i,t} + b_2 HIGH * \ln RDscale_{i,t} \\ &\quad + b_3 ENE * \ln RDscale_{i,t} + B^T X + \mu_t + \mu_i + \varpi_{i,t} \end{aligned} \tag{6.8}$$

where a and b denote the regression coefficients of the lagged values of *ESTC*.

As Eqs. (6.3), (6.4), (6.5), (6.6), (6.7) and (6.8) show, the hypotheses in this book are to verify whether the regression coefficients a_1, a_2, a_3 and b_1, b_2, b_3 in each equation are significantly greater than zero.

6.4.2 Results of Empirical Analysis

(1) Estimation Results of Mixed Data Regression

I perform regressions with Eqs. (6.3) and (6.4) by adding independent variables stepwise, resulting in estimates in Table 6.3. As shown, under different mixed data regression models, the independent variables *RDin*, ln*RDscale*, *HIGH*RDin*, *ENE*RDin*, *HIGH**ln*RDscale*, and *ENE**ln*RDscale* are all much greater than zero. Under *Model*1, which is the regression of *ESTC* on *RDin* with other control variables controlled, the regression coefficients mean that for each unit increase in R&D intensity in all industrial sectors, *ESTC* increases by 0.031. As it does not separate energy sectors and high energy-consuming sectors from industry as a whole, *Model*1 can be seen as the average effect of a one-unit increase in R&D intensity on *ESTC* in all industrial sectors. From *Model*5, one can see that a 1% increase in R&D scale corresponds to a 0.016 increase in *ESTC*. A comparison of the regression coefficients under *Model*1 and *Model*5 reveals that R&D intensity boosts *ESTC* greater than does R&D scale. *Model*2 and *Model*6 consider, based on *Model*1 and *Model*5 respectively, the effects of R&D investment in high energy-consuming sectors on *ESTC*. Under *Model*2, we can see that for each unit increase in R&D intensity in high energy-consuming sectors, *ESTC* is up by 0.248, while under *Model*6, for each 1% increase in R&D scale in high energy-consuming sectors, *ESTC* increases by 0.013. In high energy-consuming sectors, we may find, R&D intensity also drives *ESTC* faster does R&D scale. Based on *Model*1 and *Model*5, *Model*3 and *Model*7 incorporate the effects of R&D investment in energy sectors on *ESTC*. As shown in Table 6.3, in energy sectors, for a one-unit increase in R&D intensity, *ESTC* advances by 0.251; and for a 1% increase in R&D scale, *ESTC* improves by 0.031. In energy sectors, likewise, R&D intensity is more conducive to *ESTC* improvement than is R&D scale. Comparing the results under *Model*2 and *Model*6 and those under *Model*3 and *Model*7 in Table 6.3, we can find that differences in effects on *ESTC* between R&D intensity and R&D scale in high energy-consuming sectors are slightly larger than in energy sectors. That to some extent testifies to the plausibility of my treatment of R&D heterogeneity in this book. *Model*4 and *Model*8 are the full forms of mixed data regression model. As *Model*4 shows, for a one-unit increase in R&D intensity, *ESTC* increases by 0.185 on average in all industrial sectors, particularly by 0.354 in high energy-consuming sectors and by 0.41 in energy sectors; for a 1% increase in R&D scale, *ESTC* edges up by 0.015 on average in all industrial sectors, particularly by 0.025 in high energy-consuming sectors and by 0.031 in energy sectors.

Comparing *Model*1–*Model*8 leads to the revelation that both R&D intensity and R&D scale are able to accelerate *ESTC*, though R&D intensity plays a greater role than R&D scale. However, in mixed data regression models, the regression coefficients of *RDin* in *Model*1–*Model*4 vary greatly under different models relative to R&D scale coefficients, from 0.031 in *Model*1 to 0.185 in *Model*4, with the latter five times larger than the former. Although the regression coefficients of *HIGH*RDin* and *ENE*RDin* are relatively stable, the fluctuations of *RDin* regression coefficients

Table 6.3 Estimation results of mixed data regression

	Model1	*Model2*	*Model3*	*Model4*	*Model5*	*Model6*	*Model7*	*Model8*
RDin	0.031*	0.099**	0.035*	0.185*				
	(0.023)	(0.066)	(0.028)	(0.108)				
ln*RDscale*					0.016***	0.014***	0.016***	0.015***
					(0.003)	(0.004)	(0.003)	(0.004)
*HIGH**RDin*		0.149***		0.169***				
		(0.057)		(0.051)				
*HIGH**ln*RDscale*						0.009***		0.010***
						(0.002)		(0.002)
*ENE**RDin*			0.216***	0.225***				
			(0.087)	(0.084)				
*ENE**ln*RDscale*							0.015**	0.016**
							(0.009)	(0.009)
FDI	0.073*	0.071*	0.065*	0.062*	0.067*	0.069*	0.068*	0.070*
	(0.044)	(0.044)	(0.042)	(0.042)	(0.043)	(0.043)	(0.043)	(0.043)
HUM	0.787	0.788	0.686	0.683	0.792	0.757	0.848	0.757
	(0.567)	(0.568)	(0.573)	(0.573)	(0.708)	(0.723)	(0.728)	(0.745)
SOE	−0.032	−0.036	−0.036	−0.041	−0.044	−0.035	−0.030	−0.044
	(0.033)	(0.033)	(0.033)	(0.033)	(0.035)	(0.035)	(0.038)	(0.039)
ln*SPILL*	0.001	0.001	0.001	0.001	0.001	0.001	0.001	0.000
	(0.003)	(0.003)	(0.003)	(0.003)	(0.003)	(0.003)	(0.003)	(0.003)
ln*PRICE*	0.016*	0.017*	0.012	0.014*	0.017*	0.009	0.011	0.014*

(continued)

Table 6.3 (continued)

	Model1	*Model2*	*Model3*	*Model4*	*Model5*	*Model6*	*Model7*	*Model8*
	(0.010)	(0.010)	(0.009)	(0.009)	(0.010)	(0.010)	(0.009)	(0.009)
ln*CAP*	0.008	0.008	0.008	0.008	0.009*	0.009	0.008	0.009
	(0.006)	(0.007)	(0.006)	(0.007)	(0.007)	(0.008)	(0.008)	(0.008)
N	407	407	407	407	407	407	407	407
R^2	0.283	0.338	0.333	0.362	0.294	0.279	0.276	0.336

Note *, ** and *** in the above table indicate that the variables concerned are significant at the 10%, 5% and 1% significance level respectively. The values in parentheses are robust standard errors. N denotes sample size, and R^2 the adjusted goodness of fit

under different models lead to large gaps in the effects of R&D intensity on *ESTC* between high energy-consuming sectors and energy sectors. Comparing *Model*2 with Model4 and *Model*3 with *Model*4, one can see that a one-unit increase in R&D intensity would produce 0.149 and 0.169 unit of marginal effect on *ESTC* in high energy-consuming sectors, but the total effects are 0.248 and 0.354 respectively. Since the regression coefficients of *HIGH*RDin* in *Model*2 and *Model*4 are relatively stable, the regression coefficients of *RDin* contribute to the majority of the estimation deviations. The effects of R&D scale on *ESTC* are fairly stable across the econometric models. Comparisons of *Model*5 with *Model*8 and of *Model*7 with *Model*8 show that the marginal effects of each 1% increase in R&D scale on *ESTC* are 0.009 and 0.01 in high energy-consuming sectors and 0.015 and 0.016 in energy sectors.

(2) Estimation Results under Panel Data Fixed Effects Models

To the problems with the regression coefficients in *Model*1–*Model*4, a possible solution is to consider alternative forms of econometric model. I will first use the panel data regression technique to perform regressions with Eqs. (6.5) and (6.6). Since all results of the Hausman test reject panel data random effects models at least at the 5% significance level, I only report and analyze the regression results under panel data fixed effects models. See Table 6.4 for the regression results. Under different panel data fixed effects regression models, the independent variables *RDin*, ln*RDscale*, *HIGH*RDin*, *ENE*RDin*, *HIGH**ln*RDscale*, and *ENE**ln*RDscale* are positive. Except the regression coefficients of *RDin* in *Model*9 and *Model*10, all other regression coefficients are significantly greater than zero.

The regression coefficient of *RDin* in *Model*9 means that for each unit increase in R&D intensity, *ESTC* increases by 0.027 on average in all industrial sectors, controlling for other control variables and individual and time effects. Although it is slightly lower than the 0.031 in *Model*1, the regression coefficient of *RDin* in *Model*9 is not significant. The regression coefficient of ln*RDscale* in *Model*13 indicates that, controlling for other control variables and individual and time effects, a 1% increase in R&D scale would elevate *ESTC* by 0.020 on average in all sectors, an increase modestly larger than the 0.016 in the case of *Model*5. The results under *Model*10 and *Model*14 are obtained by adding R&D investment in high energy-consuming sectors to *Model*9 and *Model*13. As displayed by *Model*10's results, a one-unit increase in R&D intensity would improve *ESTC* by 0.056 on average in all industrial sectors, by 0.091 in high energy-consuming sectors in particular. Relative to *Model*2 in Table 6.3, in high energy-consuming sectors, R&D intensity has much smaller impacts on *ESTC* under the panel data fixed effects models than under the mixed data regression models. The results under *Model*14 reveal that a 1% increase in R&D scale would boost *ESTC* by 0.026 on average in all industrial sectors, particularly by 0.040 in high energy-consuming sectors. Comparing *Model*14 with *Model*6 in Table 6.3, I find that in high energy-consuming sectors, R&D scale has greater impacts on *ESTC* under the panel data fixed effects models than under the mixed data regression models.

In comparison with *Model*10 and *Model*14 in Table 6.4, *Model*11 and *Model*15 respectively equal *Model*9 and *Model*13 plus R&D investment in energy sectors. The

Table 6.4 Estimation results under panel data fixed effects models

	*Model*9	*Model*10	*Model*11	*Model*12	*Model*13	*Model*14	*Model*15	*Model*16
RDin	0.027	0.056	0.093***	0.121***				
	(0.023)	(0.055)	(0.027)	(0.058)				
ln*RDscale*					0.020***	0.026***	0.026***	0.030***
					(0.010)	(0.012)	(0.010)	(0.012)
*HIGH***RDin*		0.035***		0.036***				
		(0.017)		(0.016)				
*HIGH**ln*RDscale*						0.014**		0.010*
						(0.008)		(0.007)
*ENE***RDin*			0.233***	0.239***				
			(0.058)	(0.059)				
*ENE**ln*RDscale*							0.043***	0.041***
							(0.016)	(0.016)
FDI	0.072	0.073	0.076*	0.079*	0.067	0.067	0.035	0.037
	(0.061)	(0.061)	(0.061)	(0.061)	(0.061)	(0.061)	(0.061)	(0.061)
HUM	0.939*	0.962**	1.368***	1.423***	0.573	0.497	0.638	0.578
	(0.576)	(0.600)	(0.601)	(0.603)	(0.601)	(0.601)	(0.601)	(0.603)
SOE	−0.008	−0.005	−0.016***	−0.023***	−0.012**	−0.019***	−0.022***	−0.026***
	(0.008)	(0.008)	(0.008)	(0.007)	(0.008)	(0.009)	(0.008)	(0.009)
ln*SPILL*	0.005*	0.005*	0.005*	0.005*	0.007**	0.007**	0.007**	0.007**
	(0.004)	(0.004)	(0.004)	(0.004)	(0.004)	(0.004)	(0.004)	(0.004)
ln*PRICE*	0.000	0.001	0.006	0.007	0.019	0.019	0.018	0.018

(continued)

Table 6.4 (continued)

	*Model*9	*Model*10	*Model*11	*Model*12	*Model*13	*Model*14	*Model*15	*Model*16
	(0.031)	(0.031)	(0.031)	(0.031)	(0.032)	(0.032)	(0.032)	(0.032)
ln*CAP*	0.016*	0.017*	0.018**	0.129***	0.012**	0.011**	0.007	0.006
	(0.011)	(0.012)	(0.011)	(0.009)	(0.007)	(0.007)	(0.007)	(0.007)
N	407	407	407	407	407	407	407	407
R^2	0.129	0.177	0.153	0.288	0.126	0.131	0.193	0.309
Hausman	16.030	15.930	14.430	14.440	14.610	13.080	17.040	15.390

Note *, ** and *** in the above table indicate that the variables concerned are significant at the 10%, 5% and 1% significance level respectively. The values in parentheses are standard deviations of regression coefficients. N denotes sample size. R^2 indicates the goodness-of-fit between groups. Hausman refers to the statistics of the Hausman test under the same model

results under *Model*11 show that a one-unit increase in R&D intensity would push *ESTC* forward by 0.093 on average in all industrial sectors, by 0.326 in energy sectors in particular. In view of the regression results under *Model*3 in Table 6.3, in energy sectors, R&D intensity has larger effects on *ESTC* under the panel data fixed effects models than under the mixed data regression models. The results under *Model*15 tell us that a 1% increase in R&D scale translates to an average improvement of 0.026 in *ESTC* in all industrial sectors, 0.069 in energy sectors. Relative to the results under *Model*7 in Table 6.3, in energy sectors, R&D intensity has much larger effects on *ESTC* under the panel data fixed effects models than under the mixed data regression models.

The results under *Model*12 and *Model*16 are achieved in consequence of considering the impacts of R&D investment on *ESTC* in high energy-consuming sectors and energy sectors on the basis of *Model*9 and *Model*13 respectively. *Model*12's results indicate that a one-unit increase in R&D intensity would raise *ESTC* by 0.121 on average in all industrial sectors, by 0.157 in high energy-consuming sectors and by 0.360 in energy sectors. Compared with *Model*4's results in Table 6.3, R&D intensity has smaller effects on *ESTC* under the panel data fixed effects models than under the mixed data regression models. *Model*16's results inform that a 1% increase in R&D scale would accelerate *ESTC* by 0.030 on average in all industrial sectors, in particular by 0.040 in high energy-consuming sectors and by 0.071 in energy sectors. Compared with the regression results under *Model*8 in Table 6.3, R&D scale exerts greater effects on *ESTC* under the panel data fixed effects models than under the mixed data regression models.

(3) Estimation Results under Dynamic Panel Data Models

A common approach to regressing (the rate of) technological change is to include a lagged dependent variable in the regression model (Otto, 2006). It is significant because past technological change will influence how it develops now. On the one hand, past research on *ESTC* provides a useful foundation and tool for current research. On the other hand, as research advances, substantial breakthroughs become increasingly difficult to attain. The dependent variable *ESTC* in Eqs. (6.7) and (6.8) is the rate of technological change. Given the increasing difficulty in achieving technological change, we have reason to believe that the regression coefficients of the lagged values of the dependent variable in Eqs. (6.7) and (6.8) should be negative, which means that *ESTC* would slow down if it developed too fast in the past.

As it makes sense to use dynamic panel data models, I first estimate Eqs. (6.7) and (6.8) by the Differential Generalized Method of Moments (DGMM), and re-estimate them by the System Generalized Method of Moments (SGMM) to ensure the robustness of the estimation results (Qiang, 2010).

The results estimated by DGMM are shown in Table 6.5. As indicated by the table, the regression coefficients of the independent variables *RDin*, ln*RDscale*, *HIGH*RDin*, *ENE*RDin*, *HIGH**ln*RDscale*, and *ENE**ln*RDscale* under different models are significantly above zero, while the regression coefficients of the lagged values of the dependent variable *ESTC*(−1) are significantly below zero. *Model*17 contains the results of regression of *ESTC* on its lagged value and the independent variable *RDin*, controlling for other control variables. The regression coefficient of *ESTC*(−1) is −0.082, meaning that if *ESTC* rose by one unit in the previous period, it falls in the current period by 0.082. The regression coefficient of *RDin* is 0.756, indicating that if R&D intensity increases by one unit, *ESTC* would improve by 0.756 on average in all industrial sectors. The regression coefficients of the independent variables under *Model*21 have similar significance to those under *Model*17, except that *Model*21 represents the regression of *ESTC* on its lagged value and the independent variable ln*RDscale*, controlling for other control variables. As the regression results under *Model*21 show, if *ESTC* increased by one unit in the previous period, it declines in the current period by 0.111; if R&D scale enlarges by 1%, *ESTC* would advance by 0.037 on average in all industrial sectors.

*Model*18 and *Model*22 are derived from *Model*17 and *Model*21 respectively by picking up R&D investment in high energy-consuming sectors. As shown by the regression results under *Model*18, if *ESTC* ran faster by one unit in the previous period, it slows down by −0.089 in the current period; if R&D intensity increases by one unit, *ESTC* would increase by 0.763 on average in all industrial sectors, by 2.237 in high energy-consuming sectors in particular. Under *Model*22, a one-unit increase in *ESTC* in the previous period would drag it down by 0.118 in the current period; a 1% rise in R&D scale would speed up *ESTC* by 0.041 on average in all industrial sectors, by 0.074 in high energy-consuming sectors. *Model*19 and *Model*23 are built by adding R&D investment in energy sectors to *Model*17 and *Model*21 respectively. Under *Model*19, a one-unit increase in *ESTC* in the previous period means a 0.079 unit decrease in *ESTC* in the current period; a one-unit increase in R&D intensity would uplift ESTC by 0.786 on average in all industrial sectors, by 1.724 in energy sectors in particular. The results under *Model*23 demonstrate that a one-unit increase in *ESTC* in the previous period would be followed by a 0.108 unit fall in *ESTC* in the current period; a 1% increase in R&D intensity would boost *ESTC* by 0.039 on average in all industrial sectors, by 0.048 in energy sectors. *Model*20 and *Model*24 are constructed based on *Model*17 and *Model*21 respectively by adding R&D investment in high energy-consuming sectors and energy sectors. As shown by the results under *Model*20, if *ESTC* gained by one unit in the previous period, it would lower by 0.083 in the current period; if R&D intensity increases by one unit, *ESTC* would get faster by 0.837 on average in all industrial sectors, by 2.152 in high energy-consuming sectors. With respect to *Model*24, we see that if *ESTC* increased by one unit in the past, it would decrease by 0.115 in the current period; if R&D scale expands by 1%, *ESTC* would pick up by 0.042 on average in all sectors, by 0.073 in high energy-consuming sectors and by 0.051 in energy sectors.

As shown in Table 6.5, the regression coefficients of the main variables become more stable under dynamic panel data models when the lagged value of *ESTC* is

Table 6.5 Regression results under dynamic panel data models estimated by DGMM

	*Model*17	*Model*18	*Model*19	*Model*20	*Model*21	*Model*22	*Model*23	*Model*24
ESTC(−1)	−0.082***	−0.089***	−0.079***	−0.083***	−0.111***	−0.118***	−0.108***	−0.115***
	(0.035)	(0.034)	(0.035)	(0.035)	(0.035)	(0.035)	(0.036)	(0.036)
RDin	0.756**	0.763**	0.786**	0.837**				
	(0.448)	(0.474)	(0.500)	(0.553)				
ln*RDscale*					0.037*	0.041**	0.039*	0.042**
					(0.026)	(0.027)	(0.028)	(0.028)
*HIGH***RDin*		1.474**		1.315*				
		(0.905)		(0.932)				
*HIGH**ln*RDscale*						0.033**		0.031**
						(0.018)		(0.018)
*ENE***RDin*			0.938**	0.877*				
			(0.661)	(0.601)				
*ENE**ln*RDscale*							0.009**	0.009**
							(0.005)	(0.005)
FDI	0.479***	0.269*	0.242	0.259	0.224	0.242	0.206	0.219
	(0.219)	(0.218)	(0.220)	(0.220)	(0.225)	(0.226)	(0.230)	(0.232)
HUM	5.757***	4.990***	4.810***	4.423***	1.546	1.817	1.480	1.747
	(1.703)	(1.731)	(1.852)	(1.856)	(1.817)	(1.817)	(1.838)	(1.841)
SOE	−0.202*	−0.179*	−0.183*	−0.168*	−0.280***	−0.281***	−0.284***	−0.287***
	(0.130)	(0.131)	(0.132)	(0.132)	(0.136)	(0.136)	(0.136)	(0.136)
ln*SPILL*	0.002	0.002	0.002	0.002	0.004	0.004	0.004	0.004

(continued)

Table 6.5 (continued)

	*Model*17	*Model*18	*Model*19	*Model*20	*Model*21	*Model*22	*Model*23	*Model*24
	(0.008)	(0.008)	(0.008)	(0.008)	(0.009)	(0.009)	(0.009)	(0.009)
ln*PRICE*	0.001	0.011	0.015	0.019	0.110	0.097**	0.108***	0.095**
	(0.046)	(0.046)	(0.047)	(0.047)	(0.055)	(0.057)	(0.056)	(0.057)
ln*CAP*	0.030	0.033	0.039	0.038	0.087***	0.089***	0.085***	0.086***
	(0.033)	(0.033)	(0.034)	(0.033)	(0.034)	(0.034)	(0.035)	(0.035)
N	333	333	333	333	333	333	333	333
Chi2	35.51	37.45	35.26	37.75	17.78	18.7	17.57	18.4
Sargan	0.106	0.092	0.0805	0.224	0.143	0.132	0.128	0.112

Note *, ** and *** in the above table indicate that the variables concerned are significant at the 10%, 5% and 1% significance level respectively. The values in parentheses are standard deviations of regression coefficients. N denotes sample size. Chi2 denotes the statistics of the chi2 test, and Sargan the significance levels of the Sargan test

introduced. The influence of *ESTC* in the previous period on *ESTC* in the current period ranges from −0.079 to 0.118. On average, the range of influence on *ESTC* is 0.756–0.837 for R&D intensity, and 0.037–0.042 for R&D scale. The marginal influence on *ESTC* is 1.315–1.474 for R&D intensity in high energy-consuming sectors, 0.031–0.033 for R&D scale in high energy-consuming sectors, 0.877–0.938 for R&D intensity in energy sectors, and about 0.009 for R&D scale in energy sectors.

Considering the difficulty in estimating individual effects in Eqs. (6.7) and (6.8) by DGMM and the problem of weak instrumental variables, I have re-estimated Eqs. (6.7) and (6.8) by SGMM to ensure the robustness of the regression results. As shown in Table 6.6, the independent variables *RDin*, ln*RDscale*, *HIGH***RDin*, *ENE***RDin*, *HIGH**ln*RDscale*, and *ENE**ln*RDscale* are significantly positive under different models, while the lagged value of the dependent variable *ESTC*(−1) is significantly negative. The control variables are also consistent with the test results of the above-mentioned descriptive statistics. Relative to the DGMM results in Table 6.5, however, the regression coefficients of *RDin* are large and those of *HIGH***RDin* are small under SGMM, whereas the regression coefficients of the other independent variables under SGMM are not fundamentally different from those under DGMM.

In Table 6.6, *Model*25 and *Model*29 contain regressions of *ESTC* on its lagged value and *RDin* and ln*RDscale*, controlling for other control variables. Under *Model*25, a one-unit increase in *ESTC* in the previous period would reduces it in the current period by 0.095, and a one-unit increase in R&D intensity would improve *ESTC* by 1.102 on average in all industrial sectors. Under *Model*29, a one-unit increase in *ESTC* in the previous period would decrease it in the current period by 0.138, and 1% increase in R&D scale would boost *ESTC* by 0.048 on average in all industrial sectors. *Model*26 and *Model*30 are built by adding R&D investment in high energy-consuming sectors to *Model*25 and *Model*29 respectively. Under *Model*26, a one-unit increase in *ESTC* in the previous period would reduce it in the current period by 0.103; a one-unit increase in R&D intensity would accelerate *ESTC* by 1.293 on average in all industrial sectors, by 2.109 in high energy-consuming sectors. Under *Model*30, a one-unit increase in *ESTC* in the previous period would reduce it in the current period by 0.138, and a 1% increase in R&D intensity would raise *ESTC* by 0.047 on average in all industrial sectors, by 0.086 in high energy-consuming sectors in particular. *Model*27 and *Model*31 are built by adding R&D investment in energy sectors to *Model*25 and *Model*29 respectively. Under *Model*27, a one-unit increase in *ESTC* in the previous period would lower it in the current period by 0.099; a one-unit increase in R&D intensity would improve *ESTC* by 1.047 on average in all industrial sectors, by 1.950 in energy sectors in particular.

Table 6.6 Regression results under dynamic panel data models by SGMM

	*Model*25	*Model*26	*Model*27	*Model*28	*Model*29	*Model*30	*Model*31	*Model*32
ESTC(−1)	−0.095**	−0.103***	−0.099**	−0.105***	−0.138***	−0.138***	−0.138***	−0.137***
	(0.051)	(0.051)	(0.051)	(0.051)	(0.052)	(0.051)	(0.052)	(0.051)
RDin	1.102***	1.293**	1.047**	1.289**				
	(0.391)	(0.649)	(0.542)	(0.792)				
ln*RDscale*					0.048***	0.047***	0.047***	0.048***
					(0.022)	(0.022)	(0.022)	(0.023)
*HIGH**RDin*		0.816***		0.846***				
		(0.207)		(0.214)				
*HIGH**ln*RDscale*						0.039***		0.039***
						(0.016)		(0.017)
*ENE**RDin*			0.903**	0.871*				
			(0.513)	(0.506)				
*ENE**ln*RDscale*							0.010***	0.009***
							(0.002)	(0.002)
FDI	0.226	0.220	0.215	0.207	0.203	0.261*	0.232	0.236
	(0.173)	(0.173)	(0.191)	(0.191)	(0.189)	(0.180)	(0.215)	(0.215)
HUM	3.728***	3.064**	3.616***	3.084**	1.998	1.976	2.045	1.924
	(1.543)	(1.628)	(1.682)	(1.735)	(1.766)	(1.755)	(1.775)	(1.771)
SOE	−0.258***	−0.258***	−0.256***	−0.257***	−0.294***	−0.341***	−0.292***	−0.343***
	(0.105)	(0.104)	(0.105)	(0.105)	(0.111)	(0.112)	(0.111)	(0.113)
ln*SPILL*	0.009	0.009	0.009	0.009	0.009	0.007	0.008	0.007

(continued)

Table 6.6 (continued)

	*Model*25	*Model*26	*Model*27	*Model*28	*Model*29	*Model*30	*Model*31	*Model*32
	(0.008)	(0.008)	(0.008)	(0.008)	(0.008)	(0.008)	(0.008)	(0.008)
ln*PRICE*	0.077**	0.071**	0.076**	0.070**	0.034	0.023	0.029	0.026
	(0.027)	(0.027)	(0.027)	(0.028)	(0.027)	(0.027)	(0.030)	(0.030)
ln*CAP*	0.032	0.039	0.033	0.030	0.033	0.058**	0.033	0.057**
	(0.026)	(0.027)	(0.026)	(0.027)	(0.030)	(0.032)	(0.030)	(0.032)
N	370	370	370	370	370	370	370	370
Chi2	48.4	48.89	47.12	48.45	18.41	24.68	18.61	24.57
Sargan	0.158	0.096	0.136	0.115	0.112	0.135	0.099	0.115

Note *, ** and *** in the above table indicate that the variables concerned are significant at the 10%, 5% and 1% significance level respectively. The values in parentheses are standard deviations of regression coefficients. N denotes sample size. Chi2 denotes the statistics of the chi2 test, and Sargan the significance levels of the Sargan test

As shown by the regression results under *Model*31, a one-unit increase in *ESTC* in the previous period would reduce it in the current period by 0.138; a 1% increase in R&D scale would improve *ESTC* by 0.047 on average in all industrial sectors, by 0.057 in energy sectors. *Model*28 and *Model*32 are fully dynamic models because they incorporate R&D investment in high energy-consuming sectors and energy sectors based on *Model*25 and *Model*29 respectively. Under *Model*28, if *ESTC* increased by one unit in the previous period, it would fall by 0.105 in the current period; if R&D intensity increases by one unit, *ESTC* would rise by 1.289 on average in all industrial sectors, by 2.135 in high energy-consuming sectors and by 2.160 in energy sectors. Under *Model*32, a one-unit increase in *ESTC* in the previous period would reduce it in the current period by 0.137, and a 1% increase in R&D scale would uplift *ESTC* by 0.048 on average in all industrial sectors, 0.087 in high energy-consuming sectors and by 0.057 in energy sectors.

As shown in Tables 6.5 and 6.6, the effects of R&D intensity and R&D scale on *ESTC* are much larger under the dynamic panel models than under the mixed data regression and panel data fixed effects models, probably because the latter two types of models overlook the possible inertia of technological change.

6.4.3 Discussions of the Empirical Results

(1) Policy Implications

From the above analysis, we can come down to the following points. Firstly, all of the hypotheses to be tested are affirmed under the four types of model. Secondly, the major regression coefficients perform better under the dynamic panel models than under the mixed data regression and panel data fixed effects models. Thirdly, as the marginal effect of *RDin* on *ESTC* is greater than that of ln*RDscale*, R&D intensity has a greater boosting effect on *ESTC* than does R&D scale. Fourthly, R&D intensity has approximately the same degree of impact on *ESTC* in high energy-consuming sectors and energy sectors, and R&D scale has a larger effect on *ESTC* in high energy-consuming sectors than in energy sectors, because the (marginal and total) effects of R&D scale on *ESTC* are greater in high energy-consuming sectors than in energy sectors.

(2) Robustness Testing

Firstly, given the time series characteristics of the panel data, I use the common method of treating panel data in the literature to test whether the variables in the samples have a panel unit root, so as to ensure the robustness of the measurement results in Tables 6.4, 6.5 and 6.6 and to prevent spurious regressions. Secondly, I extend the range of samples to include those of 1990–2010, and after sorting out the samples use the new samples to re-examine the relationship between R&D heterogeneity and the rates of ESTC in China's industrial sectors.

(3) Panel Cointegration Testing

As the current first-generation panel unit root tests, such as Maddala and Wu (1999), Hadri (2000), Choi (2002), Levin et al. (2002), and Im et al. (2003), assume that the cross-sectional data (about different sectors) in the panel data are independent of each other, the test results may want robustness because the input–output linkage between different sectors is ignored. The second-generation panel unit root tests, such as Bai and Ng's (2004) PANIC and Pesaran's (2007) CIPS test, take full account of the random shocks and common factors faced by different sectors in each cross-section. For the sake of generality, I employ both the first-generation IPS test and the second-generation CIPS test to examine the stationarity of the variables used in the models. See Table 6.7 for test results.

As shown in Table 6.7, *ESTC, RDin* and ln*RDscale* are stationary series under the two panel unit root tests; *FDI,* ln*PRICE* and ln*CAP* are first order single integer series under both tests. *SOE, HUM* and ln*SPILL* are stationary series under the first-generation test, but non-stationary series under the second-generation test, probably because of the presence of large numbers of covariates among *SOE, HUM* and ln*SPILL*.

Owing to the non-stationary nature of a few variables, I further resort to panel cointegration tests to investigate whether *ESTC* has a stable long-term relationship with *RDin* and ln*RDscale*, so as to avert the spurious regression problem. The whole thrust of panel cointegration tests is to divide structurally stable heterogeneous panel data into sector-independent and sector-dependent data, and carry out cointegration tests separately. Pedroni (2004) conducted panel cointegration tests on target variables based on cross-sectional data and global data respectively and provided four test statistics. After Pedroni (2004), Westerlund (2007) has constructed a panel cointegration test under an error correction model to address possible structural mutations in panel data. As Westerlund (2007) adequately considers the possible correlations among cross-sectional data in the panel data, we may regard Westerlund's (2007) test as a modification of Pedroni (2004) (Zhiqiang, 2012). Out of these considerations, I apply Westerlund's (2007) method to test whether *ESTC* has a stable relationship with *RDin* and ln*RDscale* in the long run. See Table 6.8 for test results.

As displayed in Table 6.8, the significance levels of the panel cointegration tests of *ESTC*'s relationship with *RDin* and ln*RDscale* are not high in non-energy sectors. For example, the significance levels of Gt and Ga statistics regarding the *ESTC*-*RDin* relationship are 14.5% and 43.7% respectively, and the significance level of Ga statistics regarding the *ESTC*-ln*RDscale* relationship is 43.%%, insufficient to reject the null hypothesis that *ESTC* has is no cointegration relationship with *RDin* or ln*RDscale* in non-energy sectors. Test results from other sectors and all sectors show that *ESTC* maintains a stable cointegration relationship with *RDin* and ln*RDscale* in the long run.

Table 6.7 Results of the panel unit root test of major variables

	IPS				CIPS			
	Without time trend		With time trend		Without time trend		With time trend	
	Lag(0)	Lag(1)	Lag(0)	Lag(1)	Lag(0)	Lag(1)	Lag(0)	Lag(1)
ESTC	−11.404***	−4.440***	−9.067***	−2.872**	−2.867***	−2.314***	−2.972***	−2.342
RDin	−6.942***	−2.979***	−20.446***	4.432	−1.941*	−2.097**	−1.923	−1.564
ln*RDscale*	−9.313***	−2.614***	−22.862***	2.917	−2.473***	−1.791	−2.791***	−1.584
FDI	1.738	2.911	0.143	4.179	−1.861	−1.314	−2.619**	−1.854
SOE	−2.731***	−1.579*	0.457	−1.464*	−1.094	−0.949	−1.686	−1.296
HUM	−21.595***	−4.609***	−15.608***	−2.463***	−0.739	−0.633	−1.885	−1.631
ln*SPILL*	−21.371***	−3.303***	−25.796***	0.102	−0.604	−1.094	−2.490*	−2.267
ln*PRICE*	3.894	1.572	−0.6001	5.502	−1.859	−3.091***	−2.315	−1.966
ln*CAP*	26.428	16.953	7.162	7.971	−1.155	−1.645	−1.383	−1.482
D.*FDI*	−11.891***	−1.597*	−9.721***	−3.733***	−3.344***	−1.735	−3.352***	−2.537
D.ln*PRICE*	−15.789***	2.868	−12.572***	7.493	−2.770***	−1.913	−3.025***	−1.685
D.ln*CAP*	−2.115*	2.961	−4.626***	1.428	−1.824	−1.222	−2.256	−1.544

Note The values in the above table denote statistics under different panel unit root tests. *, ** and *** in the above table indicate that the variables concerned are significant at the 10%, 5% and 1% significance level respectively. Lag(0) and lag(1) denote no lag and a one-period lag respectively

Table 6.8 Panel cointegration test of the relationship between ESTC and R&D heterogeneity

	ESTC-RDin relationship			*ESTC*-ln*RDscale* relationship		
	Statistic	Critical value	Significance level	Statistic	Critical value	Significance level
All sectors						
Gt	−1.850	−5.111	0.000***	−2.012	−6.053	0.000***
Ga	−6.212	−3.101	0.001***	−6.098	−3.070	0.001***
Pt	−10.750	−6.550	0.000***	−10.791	−6.585	0.000***
Pa	−5.207	−8.784	0.000***	−4.503	−7.304	0.000***
High energy-consuming sectors						
Gt	−2.078	−4.492	0.000***	−2.021	−4.259	0.000***
Ga	−7.272	−3.237	0.001***	−6.523	−2.538	0.006***
Pt	−8.782	−5.669	0.000***	−8.414	−5.353	0.000***
Pa	−7.058	−8.838	0.000***	−6.054	−7.367	0.000***
Low energy-consuming sectors						
Gt	−1.635	−2.760	0.003***	−2.003	−4.302	0.000***
Ga	−5.031	−1.178	0.019**	−5.695	−1.814	0.035**
Pt	−6.991	−4.083	0.000***	−7.342	−4.384	0.000***
Pa	−4.329	−4.972	0.000***	−3.860	−4.266	0.000***
Energy sectors						
Gt	−1.890	−5.044	0.000***	−2.034	−5.843	0.000***
All sectors						
Ga	−6.359	−3.229	0.001***	−6.331	−3.194	0.001***
Pt	−10.935	−6.856	0.000***	−11.289	−7.159	0.000***
Pa	−5.671	−9.216	0.000***	−5.356	−8.591	0.000***
Non-energy sectors						
Gt	−1.526	−1.057	0.145	−1.822	−1.626	0.052*
Ga	−4.163	−0.159	0.437	−4.173	−0.163	0.435
Pt	−2.692	−1.432	0.076*	−2.744	−1.476	0.070*
Pa	−3.832	−1.938	0.026**	−2.962	−1.337	0.091*

Note *, ** and *** in the above table indicate that the variables concerned are significant at the 10%, 5% and 1% significance level respectively. The statistic, critical value and significance level in the table correspond to the value, Z-value and P-value respectively in the results of panel cointegration test xtwest in Stata. The significance levels are obtained by 500 bootstraps. The significance of Gt, Ga, Pt and Pa is given in Westerlund (2007)

(2) Tests of the Relationship between R&D Heterogeneity and ESTC in the Context of New Samples

The *China Statistical Yearbook on Science and Technology* has published statistics on internal R&D spending by sector in China for 1990–2010, but has not provided data on FDI and human capital by sector until after 2000, and *ESTC*,

RDin, and ln*RDscale* are stationary series under the two types of panel unit root test. To remain general, I deliberately drop the control variables, and re-test the relationship between R&D heterogeneity and ESTC rates with data for the 1990–2010 period by firstly processing R&D heterogeneity data for 1980–2010 by the previously described method, secondly dividing the data for the 1980–2010 period into two sets of samples, the 1980–1990 and the 1980–2010 sets of samples, and thirdly estimating the following econometric models using data from each of the two sets of samples:

$$ESTC_i = a_0 + a_1 RDin_i + a_2 HIGH * RDin_i + a_3 ENE * RDin_i + \varepsilon_i \tag{6.9}$$

$$\begin{aligned} ESTC_i = {} & b_0 + b_1 \ln RDscale_i + b_2 HIGH * \ln RDscale_i \\ & + b_3 ENE * \ln RDscale_i + \varpi_i \end{aligned} \tag{6.10}$$

$$\begin{aligned} ESTC_{i,t} = {} & a_0 + a_1 RDin_{i,t} + a_2 HIGH * RDin_{i,t} \\ & + a_3 ENE * RDin_{i,t} + \mu_t + \mu_i + \varepsilon_{i,t} \end{aligned} \tag{6.11}$$

$$\begin{aligned} ESTC_{i,t} = {} & b_0 + b_1 \ln RDscale_{i,t} + b_2 HIGH * \ln RDscale_{i,t} \\ & + b_3 ENE * \ln RDscale_{i,t} + \mu_t + \mu_i + \varpi_{i,t} \end{aligned} \tag{6.12}$$

$$\begin{aligned} ESTC_{i,t} = {} & aESTC_{i,t-1} + a_1 RDin_{i,t} + a_2 HIGH * RDin_{i,t} \\ & + a_3 ENE * RDin_{i,t} + \mu_t + \mu_i + \varepsilon_{i,t} \end{aligned} \tag{6.13}$$

$$\begin{aligned} ESTC_{i,t} = {} & bESTC_{i,t-1} + b_1 \ln RDscale_{i,t} + b_2 HIGH * \ln RDscale_{i,t} \\ & + b_3 ENE * \ln RDscale_{i,t} + \mu_t + \mu_i + \varpi_{i,t} \end{aligned} \tag{6.14}$$

The above equations represent mixed data regression, panel data fixed effects regression, and dynamic panel data regression models, respectively, with the meanings of the variables consistent with the previous sections. In this chapter, I estimate these equations separately to test the robustness of the results. See Tables 6.9 and 6.10 for estimation results.

Table 6.9 Robustness test of R&D intensity regression coefficients under different regression models

$X = RDin$		$HIGH = ENE = 0$	$HIGH = 1, ENE = 0$		$HIGH = 0, ENE = 1$		$HIGH = ENE = 1$		
		X	*X*	*HIGH*X*	*X*	*ENE*X*	*X*	*HIGH*X*	*ENE*X*
Mixed regression	1980–1990	0.041**	0.132*	0.198***	0.047**	0.287***	0.246***	0.225**	0.299***
	1980–2000	0.042***	0.136**	0.204***	0.048***	0.296***	0.253***	0.232***	0.308***
Fixed effects	1980–1990	0.033**	0.068***	0.042***	0.113***	0.282***	0.146***	0.044***	0.289***
	1980–2000	0.032**	0.067***	0.042***	0.111***	0.277***	0.144***	0.043***	0.284***
Differential GMM	1980–1990	0.809**	0.816***	1.577**	0.841***	1.004***	0.896***	1.407**	0.938***
	1980–2000	0.771**	0.778***	1.503***	0.802***	0.957***	0.854***	1.341***	0.895***
System GMM	1980–1990	1.157***	1.358***	0.857***	1.099***	0.948***	1.353***	0.888***	0.915***
	1980–2000	1.146***	1.345***	0.849***	1.089***	0.939***	1.341***	0.880***	0.906***

Note *, ** and *** in the above table indicate that the variables concerned are significant at the 10%, 5% and 1% significance level respectively

Table 6.10 Robustness test of R&D scale regression coefficients under different regression models

$X = \ln RDscale$		$HIGH = ENE = 0$	$HIGH = 1, ENE = 0$		$HIGH = 0, ENE = 1$		$HIGH = ENE = 1$		
		X	X	$HIGH*X$	X	$ENE*X$	X	$HIGH*X$	$ENE*X$
Mixed regression	1980–1990	0.021***	0.019***	0.012***	0.021***	0.020***	0.020***	0.013***	0.021***
	1980–2000	0.022***	0.019***	0.012***	0.022***	0.021***	0.021***	0.014***	0.022***
Fixed effects	1980–1990	0.024**	0.031***	0.017***	0.031***	0.052***	0.036***	0.012***	0.050***
	1980–2000	0.024**	0.031***	0.017***	0.031***	0.051***	0.036***	0.012***	0.049***
Differential GMM	1980–1990	0.040***	0.044**	0.035**	0.042***	0.010***	0.045***	0.033**	0.010***
	1980–2000	0.038***	0.042***	0.034***	0.040***	0.009***	0.043***	0.032***	0.009***
System GMM	1980–1990	0.050***	0.049***	0.041***	0.049***	0.011***	0.050***	0.041***	0.009***
	1980–2000	0.050***	0.049***	0.041***	0.049***	0.010***	0.050***	0.041***	0.009***

Note *, ** and *** in the above table indicate that the variables concerned are significant at the 10%, 5% and 1% significance level respectively

References

Ahmad, S. (1966). On the theory of induced invention. *The Economic Journal, 76*(302), 344–357.

Arrow K. J. (1962). The economic implication of learning by doing. *The Review of Economic Studies*.

Bai, J., & Ng, S. (2004). A PANIC attack on unit root and cointegration. *Econometrica, 72*(4), 1127–1178.

Barro, R. J. (2001). Human capital and growth. *American Economic Review, 91*(2), 12–17.

Caselli, F., & Coleman, W. J. (2001). Cross-country technology diffusion: The case of computer. *American Economic Review, 91*(2), 328–335.

Chengze, F., Yifan, H., & Hongliang, Z. (2008). A theoretical and empirical study on the impact of FDI on domestic firms' technological innovation. *Economic Research Journal, 01*, 89–102.

Chi, M., Changlin, G., & Han, S. (2003a). Energy R&D investment in China. *Energy Research & Utilization, 03*, 3–6+38–48.

Chi, M., Changlin, G., & Han, S. (2003b). A study of energy R&D investment in China. *International Petroleum Economics, 04*, 37–40+64.

Chi, M. (2000). Imbalanced energy R&D in China's R&D investment structure. *East China Science & Technology, 01*, 32.

Choi, I. (2002). Unit root tests for panel data. *Journal of International Money and Finance, 20*(2), 249–272.

Cohen W. M., & Levinthal D. A. (1989). Innovation and Learning: The Two Faces of R&D. *The Economic Journal*.

Engelbrecht, H. J. (1997). International R&D spillovers, human capital and productivity in OECD economies: An empirical investigation. *European Economic Review, 41*(8), 1479–1488.

Fan, C., & Hu, Y. F. (2007). Foreign direct investment and indigenous technological efforts: Evidence from China. *Economic Letter, 96*(2), 253–258.

Genfu, F., Junhu, L., & Zhilin, X. (2006). An empirical analysis of R&D efficiency and its influence factors in China's industrial sectors. *China Industrial Economics, 11*, 46–51.

Hadri, K. (2000). Testing for stationarity in heterogeneous panel data. *The Econometrics Journal, 3*(2), 148–161.

Haiyang, Z. (2008). The impact of FDI on China's industrial innovation capability—The determinants of independent innovation. *Journal of International Trade, 01*, 72–81.

Hicks, J. R. (1932). *The Theory of Wages[M]*. Macmillan.

Hongfeng, Z., & Zhaozhong, Z. (2013). The impact of ownership structure reform on total factor productivity in industrial sectors—From the perspective of easing entry regulation. *Economic Theory and Business Management, 02*, 66–77.

Hongling, W., Daokui, L., & Junxin, F. (2006). FDI and independent R&D: An empirical study based on industry data. *Economic Research Journal, 02*, 44–56.

Im, K. S., Pesaran, M. H., & Shin, Y. (2003). Testing for unit roots in heterogeneous panels. *Journal of Econometrics, 115*(1), 53–74.

International Energy Agency. (2011). *IEA Guide to Reporting Energy RD&D Budget/Expenditure Statistics*. IEA. Retrieved from http://www.iea.org/stats/RDD%20Manual.pdf

Kouvaritakis N., Soria A., & Isoard S. (2000). Modelling energy technology dynamics: Methodology for adaptive expectations models with learning by doing and learning by searching. *International Journal of Global Energy Issues*.

Levin, A., Lin, C. F., & Chu, C. S. J. (2002). Unit root test in panel data: Asymptotic and finite sample properties. *Journal of Econometrics, 108*(1), 1–24.

Liangke, X. (2010). How human capital and R&D affect total factor productivity—An empirical analysis based on large and medium-sized industrial enterprises in China. *The Journal of Quantitative & Technical Economics, 04*, 78–94.

Maddala, G. S., & Wu, S. (1999). A comparative study of unit root tests with panel data and a new simple test. *Oxford Bulletin of Economics and Statistics*, 631–652.

Otto, V. M. (2006). *Economic modeling of directed technical change: The case of CO_2 emission reduction*. PhD dissertation, Wageningen University.

Pedroni, P. (2004). Panel cointegration: Asymptotic and finite sample properties of pooled time series tests with an application to the PPP hypothesis. *Econometric Theory, 20*(3), 597–625.

Pesaran, M. H. (2007). A simple unit root test in the presence of cross-section dependence. *Applied Econometrics, 22*(2), 265–312.

Qiang, C. (2010). *Advanced Econometrics and Stata Applications*. Higher Education Press.

Ran, W., Bo, Y., & Weigen, D. (2010). How FDI influences China's industrial innovation capability—From the perspective of industrial linkages. *China Industrial Economics, 11*, 16–25.

Rosenberg, N. (1982). *Inside the Black Box: Technology and Economics*. Cambridge: Cambridge University Press.

Wenjun, Z., & Jinping, Y. (2012). Trade openness, FDI and China's industrial growth pattern: An empirical study based on data from 30 industrial sectors. *Economic Research Journal, 08*, 18–31.

Westerlund, J. (2007). Testing for error correction in panel data. *Oxford Bulletin of Economics and Statistics, 69*(6), 709–748.

Xian, Z., & Yong, Z. (2007). The spatial effect of FDI on energy intensity in China. *The Journal of Quantitative & Technical Economics, 01*, 101–108.

Xuan, W., Zhengzheng, B., & Jiuli, H. (2012). Inter-sectoral FDI spillovers and provincial industrial growth. *South China Journal of Economics, 05*, 16–26.

Yu, C. & Guoliang, K. (2009). Market structure and FDI technology spillover—An empirical study based on dynamic panel data of Chinese manufacturing industry. *World Economy Studies, 09*, 63–68+89.

Zhengge, T., & Geng, X. (2006). Transforming China's industrial growth model—A nonparametric production frontier dynamic analysis of labor productivity in large and medium-sized enterprises. *Management World, 10*, 57–67+81.

Zhigang, W., Liutang, G., & Yuyu, C. (2006). Decomposition of inter-regional productivity and TFP growth (1978–2003). *Social Sciences in China, 02*, 55–66+206.

Zhiqiang, Z. (2012). Financial development, R&D innovation and regional technology deepening. *Economic Review, 03*, 82–92.

Chapter 7
Main Findings and Research Outlook

7.1 Main Findings

Based on the existing theories and analytical methods, I have attempted to clarify and define the basic concept and connotations of energy-saving technological change, distinguish between energy R&D investment and non-energy R&D investment, and explore how R&D heterogeneity affects ESTC, and incorporate two effects of R&D heterogeneity on energy efficiency, i.e., the effect of factor substitution and the effect of increased effective energy input, into one analytical framework. By doing so, I have for the first time built an analytical framework for endogenous ESTC in China's industrial sectors from the perspective of R&D heterogeneity, thus expanding the existing analytical framework for endogenous ESTC.

A vast and detailed literature review shows that exogenous ESTC theory, represented by the DHSS model, seems to be waning, while the research on endogenous ESTC characterized by its endogenous nature, is thriving. As to what results the research has accomplished, however, no consensus has been reached, as there is still large room for improvement—even the basic concept and meanings of ESTC are controversial. Therefore, it is extremely worthwhile to precisely define ESTC, explore how R&D heterogeneity possibly affects ESTC, and expand the existing frameworks of analysis for endogenous ESTC.

The existing literature on endogenous ESTC—both endogenous embodied technological change theory and R&D-induced technological change theory—emphasizes the effect of substitution of better non-energy production factors (especially capital) and knowledge accumulation for energy efficiency, and largely ignores the possible impact of increased effective energy input on energy efficiency; few studies have attempted to incorporate the effects of factor substitution and increased effective energy input in one analytical framework. Although as early as the 1980s Solow (1974), and Dasgupta and Heal (1974) highlighted factor substitution and increased effective energy input as the keys to breaking growth limitations and guaranteeing sustainable growth, I have found few studies in the literature that consider both effects in one analysis. Hassler et al. (2012) have partially redressed the problem by assuming

X. He, *The Endogenous Energy-Saving Technological Change in China's Industrial Sector*, https://doi.org/10.1007/978-981-19-7485-4_7

energy-augmenting technological change in the economy, and have explored the effect of increased effective energy input on energy efficiency; their model based on R&D heterogeneity has provided useful insights for this book. However, Hassler et al.'s (2012) assumption that capital and energy are fully complementary to each other greatly limits the applicability of their model. For this reason, I relax the key assumptions in the existing models, and by allowing for the possibility of substitution between factors of production, and distinguishing between energy R&D investment (investment that increases effective energy input by producing ESTC) and non-energy R&D investment (investment that improves the quality of capital goods, reducing the amount of energy used in production), build a self-consistent logical chain from R&D heterogeneity to energy efficiency improvement to sustainable growth.

Based on the theoretical analysis, I further estimate the rates of ESTC in China's 37 industrial sectors from 1980 to 2010 and empirically examine the relationship between R&D heterogeneity and ESTC. The difficulties in the empirical study include: first, how to estimate the three-factor (including energy) nested CES production function; second, how to address the problem of inconsistent statistical and classification standards relating to Chinese industry; and third, how to properly measure R&D heterogeneity.

In summary, this book purports to build a framework for analyzing endogenous ESTC and propose the possible roles of R&D heterogeneity in breaking the energy and carbon dioxide emission constraints, decarbonizing Chinese industry, and achieving sustainable growth. I have explored the mechanisms whereby R&D heterogeneity may improve energy efficiency in China's industrial sectors, distinguishing between the effect of factor substitution and the effect of increased effective energy input in relation to R&D heterogeneity's impact on energy efficiency.

7.2 Policy Recommendations

My point of departure is to explore what impacts R&D heterogeneity may have on ESTC and how. I expect the main findings in this book to be applicable not only to China's industrial sectors, but to other economic actors as well. In fact, the issue of whether to invest R&D in energy sectors or in non-energy sectors faces all economic actors, so long as they need energy in their production processes, have R&D departments, and have to make decisions about R&D investment. The framework of analysis constructed in this book, therefore, is largely applicable to all of them. Based on the findings in this book, I propose the following policy recommendations.

Firstly, at the macro-level, since energy R&D investment can effectively stimulate ESTC, subsidizing energy R&D investment in energy sectors and high energy-consuming sectors may encourage them to invest in energy R&D, driving long-term sustainable growth and decarbonization of China's industrial sectors. In addition, given the heterogeneous nature of R&D investment, policy makers should, when designing subsidy policies, consider the impacts of R&D investment on energy and emissions reduction efforts from both the factor substitution and the increased effective energy input perspectives, and should appreciate the differences and connections between different energy efficiency metrics. If we only focus on the factor substitution effect, or make no distinction between different energy efficiency metrics, we will not only misunderstand how R&D investment operates to reduce energy consumption and carbon emissions, but also underestimate the effects of subsidy policies.

Secondly, at the sectoral level, we should adjust and improve the subsidy policy for energy R&D investment in accordance with the characteristics of different sectors. As the findings in this book suggest, while greater R&D intensity and R&D scale can promote ESTC in both energy sectors and high energy-consuming sectors, R&D scale is more significant in high energy-consuming sectors. In designing the R&D subsidy policy, therefore, policy makers should aim at increasing R&D intensity for energy sectors, while enlarging R&D scale for high energy-consuming sectors (see Chap. 6 for the definitions of energy sectors and high energy-consuming sectors), so as to promote ESTC effectively in China's industrial sectors. What's more, we should monitor how the R&D subsidy policy is implemented. Not only should we focus on increases in R&D intensity or R&D scale in a certain sector, but also know where R&D investment goes in that sector.

Thirdly, at the factor of production level, we should improve market mechanisms for energy as soon as possible. As concluded in Chap. 4, the scarcity of energy sources would increasingly drive R&D investment into ESTC research. But that conclusion would lie only if that scarcity could be truly reflected in energy prices. It is imperative, therefore, to build and improve energy markets, where energy prices could reflect the scarcity of energy, thus incentivizing energy-consuming sectors to engage in R&D activities for ESTC.

7.3 Research Outlook

In explaining ESTC, different researchers may come to different conclusions as the methods, underlying assumptions and theories they employ vary, limiting in some degree the applicability of each theory. This book has tracked and studied the representative studies and latest results of (heterogeneous) R&D investment-based endogenous ESTC in China and other countries, and has made useful explorations in several aspects. Owing to various subjective and objective limitations, this book is not free from some shortcomings as follows.

(1) With regard to the endogenous mechanism whereby ESTC is possible, I mainly discuss how "endogenous" progress in energy-saving technologies is achieved, at least including through R&D and learning-by-doing. I only explore it at the R&D level, leaving other aspects untouched. As the globalization of the Chinese economy deepens, China's industrial sectors will get more and more involved in the international division of labor; one possible focus of future research is to

understand how learning-by-doing based on the division of labor affects ESTC and what the transmission mechanism is.

(2) In respect of modeling, I first assume that the elasticity of substitution between factors of production is fixed at 1, and then relax this assumption to focus on the possible effects of the elasticity of substitution between energy and capital on equilibrium ESTC and economic growth. This method can moderately simplify the analysis process and lead to a series of conclusions, which may be controversial, yet it impedes me from fully analyzing the quantitative relationship between changes in the elasticity of substitution and ESTC. Extending the subject of this book to include all multi-factor CES nested production functions, or even Varying Elasticity of Substitution (VES) production functions, and arriving at all equilibrium solutions, may be the key of future research. Few studies have been carried out on the theory of endogenous technological change based on CES production functions; many researchers have extended the Solow model and the basic Ramsey model through CES production functions. These models are hardly competent to produce analytical equilibrium solutions, and researchers often rely on huge quantities of numerical simulations to perform equilibrium approximation. Future research, therefore, is as important as the introduction of CES production functions or even VES production functions to the theory of endogenous technological change, and will become a powerful analytical tool comparable with those two types of production function.

(3) In terms of underlying statistics, though it is reasonable to determine R&D heterogeneity by the nature of a sector and energy consumption levels, and to measure it by R&D spending in a given sector, it would be helpful to future research to improve the existing statistics or to conduct wholesale investigations of certain major sectors, in order to distinguish between energy R&D investment and non-energy R&D investment in different sectors, or even go beyond the dichotomy used in this book and extend heterogeneity to multiple types of R&D investment to make the statistics more accurate and relevant. China's industrial sectors have long been plagued with inconsistent statistical and classification standards, which may explain why the relevant research has fallen behind international levels. However, this has provided researchers with a new task of research. In the future, we should tinker with the existing indicator system, accumulate more data and information, and build a better, detailed and accurate data system.

(4) In the selection of empirical models, the three mainstream international estimation approaches all disregard the correlations between different sectors, and this book is no exception. In other words, in studying the three-factor nested CES production function applicable to industrial sectors, the researchers estimate each sector separately. How to consider the correlation between different sectors is a pressing issue for all current CES production function estimation approaches. The basis of estimation in this chapter is the CES production function, which greatly simplifies the analysis process by assuming constant elasticity of substitution. While in a real-world economic environment the elasticity of substitution may remain constant in the short run, long-term economic

data do not support this conclusion (de La Grandville, 2009). As Chinese industry has grown leaps and bounds since the 1980s, it is hardly convincing to continue to assume a constant elasticity of substitution. Developing an estimation approach that considers both changes in the elasticity of substitution and sectoral correlations, therefore, is an issue calling for solution.

References

Dasgupta, P., & Heal, G. (1974). The optimal depletion of exhaustible resources. *Review of Economic Studies, 41*(5), 3–28.

de la Grandville (2009). *Economic Growth: A Unified Theory*. Cambridge: Cambridge University Press.

Hassler, J., Krusell, P., & Olovsson, C. (2012). *Energy-saving technical change* (NBER Working Paper No. 18456).

Solow, R. M. (1974). Intergenerational equity and exhaustible resources. *Review of Economic Studies, 41*(5), 29–45.

The manufacturer's authorised representative in the EU is Springer Nature Customer Service Centre GmbH, Europaplatz 3, 69115 Heidelberg, Germany. If you have any concerns regarding our products, please contact ProductSafety@springernature.com

Printed and bound by CPI Group (UK) Ltd, Croydon, CR0 4YY
30/09/2026
02225259-0017